G. P. Chapman and Y. Z. Wang

The Plant Life of China

Springer

Berlin
Heidelberg
New York
Barcelona
Hong Kong
London
Milan
Paris
Tokyo

Geoffrey P. Chapman · Yin-Zheng Wang

The Plant Life of China

Diversity and Distribution

With 23 Figures, Including 16 Colour Plates,
and 44 Tables

 Springer

Dr. Geoffrey P. Chapman
Mistral
3 Oxenturn Road
Wye, Ashford
Kent TN25 5BH
UK

Professor Dr. Yin-Zheng Wang
Institute of Botany, the Chinese Academy of Sciences
20 Nanxincun, Xiangshan
Beijing, 100093
P. R. China

Cataloging-in-Publication Data applied for
Die Deutsche Bibliothek – CIP-Einheitsaufnahme

Chapman, Geoffrey P.:
The plant life of China : diversity and distribution / Geoffrey P. Chapman and Yin-Zheng Wang. -
Berlin ; Heidelberg ; New York ; Barcelona ; Hong Kong ; London ; Milan ; Paris ; Tokyo : Springer,
2002
   ISBN 3-540-42257-9

ISBN 3-540-42257-9 Springer-Verlag Berlin Heidelberg New York

Springer-Verlag Berlin Heidelberg New York
a member of BertelsmannSpringer Science+Business Media GmbH

http://www.springer.de

© Springer-Verlag Berlin Heidelberg 2002
Printed in Germany

Production: PRO EDIT GmbH, 69126 Heidelberg
Cover Design: design & production GmbH, Heidelberg
SPIN: 10783309   31/3130/hs   5 4 3 2 1 0 – Printed on acid-free paper

I do not enlighten anyone who is not eager to learn, nor encourage anyone who is not anxious to put his ideas into words.

Kong fu zi 551 – 479 B. C.

(Confucius)

Acknowledgements

To assemble a book of this nature depends not only on the resources of various institutes but on the cooperation and goodwill of many experts, and it is a pleasure to thank them for time, interest and skill given so readily. In particular we wish to thank the following in China: Ma Keeping, Qin Hai-Ning, Wang Siyu and Wang Won Tsai (CAS Institute of Botany, Beijing); Zan Xinglan (National Natural Science Foundation of China); Paul Wusteman (British Embassy); Lui Mei and Yu Ge (China EU Centre for Agricultural Technology, Beijing); Ren Jizhou, Nan Zhi Biao, Wang Yan Rong and Zhang Zihe (Grassland and Ecological Research Station, Gansu); Wu Qi Gen and Lui Lin (CAS South China Institute of Botany, Guangdong); Hu Yaohua (South China Academy of Tropical Crops, Hainan); Richard Corlett, John Hodgekiss (University of Hong Kong) and Gloria Sui Lai-Ping, (Kadoorie Foundation, Hong Kong); Chen Shou-liang, (Jiangsu Institute of Botany, Nanjing, Jiangsu); Yuying Gang (CAS Hauxi Sub alpine botanic garden); Zhang Xin Yue (Station for Grassland Management, Sichuan); Li Dezhu, Li Heng and Wang Hong (CAS Institute of Botany, Kunming) and Wang Kanglin, Xue Ji Rue and Yang Yuming (South Western Forestry College, Kunming, Yunnan).

In the West, we record our thanks to Jeff Moorby (Annals of Botany Company); Steve Vincent (Cambridge Arctic Shelf Programme); David Chamberlain and David Watson (Royal Botanic Gardens, Edinburgh); Aljos Fargon; Christine Leon, Simon Owen and Steve Renvoize (Royal Botanic Garden, Kew); Marquita Baird and John Marsden (Linnean Society of London); Paul Kendrick (Natural History Museum, London); John Moffat (Needham Research Institute, Cambridge); Michael Boulter and Andrew Roberts (University of East London) David Phillipson and Elizabeth Williamson (School of Pharmacy, University of London) and Ping Jin (Wye College)

Grateful indeed as we have been for help, responsibility for the content of the book, rests of course, with the authors.

Our thanks are also due to Jocelyn Hart, who finalised the camera-ready copy. Finally, we thank our wives, who have made every allowance for us.

Preface

This book is for anyone interested in plants, whether they live inside or outside of China. We could, as has been done before, have produced a book crammed with colour photographs so as to delight the eye – but in so doing numb the mind. Manifestly, that is not our intention. Rather, we have used China and its flora so as to concentrate upon the botanical issues involved. We try to explain how it is that China is botanically so rich and how, too, this richness is not spread generally but localised in particular areas. To do this, it is necessary to explore, if only briefly, the geological and climatic history of China.

Those of us schooled in a Western tradition are aware of a Greek and Roman inheritance followed by the Dark Ages when, for a millenium and more following Dioscorides, intellectual interest in plants remained dormant almost until the Renaissance. In China, the situation was quite different. There is a botanical tradition here that predates Theophrastus and continued, almost uninterruptedly, until it merged with that brought to China from the West about 400 years ago – significantly predating Linnaeus.

Like a magnet, the flora of China has drawn famous plant collectors such as Ernest Wilson, George Forrest and Frank Kingdon-Ward, and there is no doubt as to how much gardens around the world owe to their efforts and those of others less well known. How rich, though, is the Chinese flora? A part of this book considers that question.

At first acquaintance, China's appearance can be, botanically, misleading. To visit the busy countryside is to see agriculture stretching to the horizon. In the cities, amid vast conurbations, is that huge bicycling multitude now increasingly adopting cars. It is true that on apartment balconies one can find plants perched in pots and around some of the newer building developments there are amenity plantings but, one might ask "where are the plants for which China is so justly famous?" This is another of the questions this book seeks to answer.

Much more so than in most Western countries, there is in China a strong medicinal plant tradition. How significant is it? Is it inevitably to be replaced by Western medicine or is it a viable intellectual tradition in its own right before which we should pause and find what it has to teach us? Here, too, we explore a contrast between Western and Oriental traditions.

Anyone outside of China, especially if they see themselves as botanists rather than linguists, could well be diverted from studying the Chinese flora by the language barrier. The matter is, for several reasons, less formidable than it might appear. Increasingly, Chinese scientists write and publish in English. Again, Chinese botanical texts often use Latin plant names and, even without fluency in

Chinese, such texts can be helpful. We indicate in this book various sources of information accessible to English speakers.

At institutions throughout China botanists are at work collecting, sorting, mounting, labelling and assessing the immense flora. It is a peaceable and necessary work among what are priceless repositories of plant material, and is nowadays complemented by the newer disciplines of DNA screening, for example. New and old have a place, and we attempt here an account of how effort in various directions is distributed.

In a book of this size we cannot attempt to cover everything which preoccupies us. We might hope, however, to achieve three objectives. These are to share what we have to offer, to engage the reader's attention and perhaps, too, to whet his or her appetite for a flora of compelling interest and global significance.

G. P. Chapman

[1]Wang Yinzheng

---

[1] Supported by the State Key Basic Research and Development Plan (G2000046803-4)

Contents

Part I – The Setting

Part II – Growth Forms: Some Representative Taxa

Part III – Some Major Genera

Part IV – Conservation and the Environment

Appendices

References

A Note on Chinese Script and Language

Chinese script is not alphabetic. Various schemes have been devised to represent written and spoken Chinese in the 26 letters of the Latin alphabet, based on the pronunciation of the Peking dialect of spoken Chinese, or 'mandarin'. The people's Republic of China has adopted 'pinyin' romanisation as such a standard, and this has been used in this book for Chinese names and places. We have, however, made an exception in the case of references to classical Chinese literature derived from the works of Joseph Needham and his associates. Here, their amended 'Wade-Giles' romanisation is retained to simplify reference to those sources.

A Note on Plant Family Names

Alternate family names are a fact, however inconvenient, with which botanists have to deal. The approach adopted here is to accept the original author's usage, i.e. Asteraceae/Compositae, Apiaceae/Umbelliferae, Gramineae/Poaceae, Fabaceae/Leguminosae, Labiatae/Lamiaceae, Palmae/Arecaccae and to indicate the equivalencies at convenient points in the text.

Part I

The Setting

1 An Introduction to Chinese Plant Science

Furthermore, the best Chinese agricultural treatises, in our opinion, surpass anything produced before the 18th century in Europe in that systematic presentation of technical detail. Only in the treatment of such specialist topics as viticulture can the great works of Varro and Columella compare with the *Chhi Min Yao Shu* and its successors for pragmatic details. The Chinese works cover a much wider range of crop plants than Western works, including not only numerous varieties of cereals and fibre crops but also vegetables, fruits, citrus, sugar cane and tea. The number of species, and also of varieties, recorded and described in Chinese agronomic literature is incomparably greater than anything to be found in the pre-modern West.

Bray (1984)

Anyone who knows Chinese literature is aware that there was not a single century between A.D. 100 and 1700 that did not see the appearance of at least one new and original work on pharmaceutical natural history, and some which saw many.... in other words Chinese botany had no Renaissance and it had no "dark ages" either

Lu and Huang (1986)

Any Westerner seeking to understand Chinese science needs to recognise two facts in particular. The first is the prolific output of Chinese literature which occurred over two millennia. Secondly that a Cambridge scholar Joseph Needham (1900–1995), has organised a truly monumental assessment of this science in the multivolume *Science and Civilisation in China*. The first volume appeared in 1954 and the eighth is the most recent to appear in 1999. Volume is perhaps a deceptively understated word. The whole venture is a collaborative one involving other contributors but the driving force for the whole project was generated by Needham and is breathtaking in its depth and breadth. Not surprisingly, the résumé offered in this chapter is largely derived from Volume 6, Biology and Biological Technology. Given the importance of this volume in the present context, the wider project of which it is a part and one or two unusual aspects of it, the opportunity is taken at this point to introduce it to the reader. It is in three parts whose dates of publication do not follow the expected sequence. Part 1 (1986) by Lu and Huang deals with Botany. Part 2 (1984), by Bray concerns Agriculture. Part 3 (1996), Agro-Industry is sub-divided into Agro-industries: Sugar Cane Technology by Daniels and Forestry by Menzies. It is important to point out that each part contains three types of references – A Chinese and Japanese books before 1800 A. D.; B Chinese and Japanese books and journal articles since 1800 and C, books and articles in Western languages. The coverage of literature can therefore be over two or three millenia.

Earliest Traditions

From the misty origins of any civilisation there is distilled a residue of folklore and eventually literature. For comparison, the epic Greek poet Homer who composed the *Odyssey* BC and the *Iliad* can hardly have been active before the 8th century. In China however the Book of Songs records folk poems and songs circulating among the people during the Zhou Dynasty from the 11th to the 6th century B. C. Confucius finally revised the book and there are named here more than 100 species of plants.

The Zhou Li (*The Practice of the Zhou Dynasty*) is a 4th B.C. production where written characters for animal and plant are beginning to be stabilised. For references see Wang (1991), Wang (1994), Gou (1989) and Xia (1981).

From Kong-fu-zi (Confucius) 551-479 B.C.

The Chinese intellectual tradition dates largely from Confucius, whose more immediate followers assembled the *Analects,* a collection of sayings ascribed to him. From there, divergent streams characterised Chinese culture and generated an immense literature. It should be recognised that the rudiments of printing were available in the 9th century A.D. and that with the Song Dynasty thereafter in the 10th century, printing was well established, thus anticipating a similar development in Europe by about 400 years.

European plant science was begun by the Greeks and then, in a more applied way, adopted by the Romans. Thereafter a whole millennium elapsed before genuinely new work was undertaken. During the corresponding time in China numerous works emerged, many of them still extant and offering a startling contrast between the Western and Oriental traditions. Not every Chinese work was original by any means, but the contrast is there and is not readily explained.

While languages can and do change through time, the written characters in which they are expressed can be surprisingly durable. Two examples from China are the characters for animal and plant.

	3rd Century BC	Modern Chinese	Pronunciation
Animal	動物	动物	dong wù
Plant	植物	植物	zhí wù

East · West Contact

While full contact between the Old World East and West cultures was not established until after 1800, there had previously been occasional exchanges and these raise questions about how far the one stimulated the other. Alexander the Great had reached the Indus in 326 B.C. and Theophrastus, in compiling his taxonomy was aware of plants which had been brought to Greece as a result of Alexander's travels. Philo (15 B.C.–47 A.D.) mentions a school of Indian philosophers, the Gymnosophists.

The principal line of contact was the Silk Road, beginning in Xian, following the Great Wall and then going across Central Asia to the Levant. From Rome eastward passed wool, gold and silver and from China westward, silk. One might assume ideas would travel reciprocally, but mostly goods were moved by a series of middlemen, primarily traders. With the collapse of Rome and the eventual rise of the Arab civilisation, travel was more precarious and diminished, though not before paper had been brought from the East to the West in the 12th century.

Clement of Alexandria (200 A.D.) indicated some awareness of Buddhism and that Nestorian Christians from the West co-existed with Chinese Buddhists, a fact commemorated on an inscription in Xsianfu from 781 A.D., (Kung 1987). There appear to have been Jewish merchants trading between China and Provence from the beginning of the 9th century AD[1]. Before 1552 the Jesuit, Francis Xavier, had formed a friendship with a Buddhist abbot in Japan (Kung 1987). Quite apart from any religious dialogue, one assumes there could have been a reciprocal interest in medical traditions and the use of herbal remedies. Later, the Jesuit missionaries formed an important East–West link. They took with them a knowledge of the strongly developed physical science from Europe but botanically they were mostly recipients of the riches which China had to offer. In this, as in other disciplines, their commitment was impressive and their contribution is commemorated in a long list of Latin binomials. One significant exception to this is *Cinchona ledgeriana,* an antimalarial plant from the New World, a source of quinine, known by its common name as Jesuits bark brought by them to China in the 15th century. Interestingly, of all the Old World cultures, only the Chinese had discovered an antimalarial remedy, *Dichroa febrifuga,* and about this they had been aware since the 3rd century B.C.

Table 1.1 indicates something of the contrast between China and Europe in terms of plant-related output, but the former references are merely a sample and far from exhaustive. One could readily augment the table by pointing out that

---

[1] Marco Polo (1254–1324) has a mixed reputation. Having apparently spent 17 years in China he remained oblivious of printing. Bretschneider, (1898) comments that Marco Polo was seemingly unaware of tea drinking in China.

numerous pharmacopoeias[2] were produced before 1800 together with publications committed to particular plant groups. For example, between 560 and 1670 there were five publications on bamboo, from 1186 and 1352, three on apricots, from 770 and 1430 seven on tea, between 1104 and 1798 no fewer that 15 on *Chrysanthemum* and between 986 and 1793 a remarkable 18 on tree-peonies.

Monographs

Among the surprises is the fact that the Chinese began writing plant monographs so early. The *Chu Phu* of 460 A.D. or thereabouts on bamboo appears to be the earliest monograph of any plant group in any culture. It is partly in the form of poems and recognises bamboos as "neither herb nor tree". There is considerable awareness of morphological detail, habitat preference and holocarpy. There is too a distinction, inferred between what we now call pachymorph and leptomorph rhizomes. Subsequent treatises on bamboo appeared in 970, 1299, the 14th century and in 1670. That of 1299 has impressively detailed illustrations.

Although several other groups might be used to provide monograph examples, this section concludes by mentioning the *Chü Lu* (*The Orange Record*) of 1178. Needham et al. (1986) refer to its "28 genera, species, sub-species and varieties described by Han Yen-chic" and then seek to relate these to a modern Latin binomial assessment of the Aurantioideae, notwithstanding that taxonomic opinion differs regarding this problematic group of plants.

The Pharmacopoeia

Chinese interest in herbal remedies is well known and, as indicated earlier, has a long history. This matter is reserved for a later chapter.

Famine Plants

One curious and humane aspect of Chinese plant interest has been the search for species to which people might turn in times of famine – a trend Needham et al (1986) called the Esculentist Movement. Since some of the plants selected were poisonous, or at least disagreeable, part of the concern lay in how to detoxify

---

[2] With what seems to be in line with Chinese humour, one pharmacopoeia title is translated as *The Pharmacopoeia Purged* (779 A.D.) and another as *Needles from the Haystack* (1668 A.D.)

Table 1. 1. Contrast between China and Europe in terms of plant-related output

	China	Europe
11th- 6th C.	*The Book of Songs*	
4th C.	*The Practice of the Zhou Dynasty*	
3rd C.	*The Book of Master Kuan* (late 4th C, contains information on soils.	Theophrastus - *Enquiry into Plants*
	Erh Ya - Literary Expositor.	
	Analects	
2nd C.		Cato – *De agri culture*
1st C. B.C.	Commentary on the text of the *Pharmacopoeia of the* [3]*Heavenly Husband-man.*	Varro – *Res rustica*
		Columella – *De de rustica, De arborilus*
1st C. A.D.	*The Book of Fan Shêng-Chih Shu.* (on Agriculture).	Virgil – *Georgics*
2nd C.		Pliny the Elder *Historia naturalis*
		Dioscorides – *Materia Medica*

[3] According to a Chinese myth of rather late origin, Shen Nung, the Heavenly Husbandman was a Sage-emperor who lived in the 3rd millennium B.C. and taught his people the use of the plough and the cultivation of cereals (Bray 1984).

Table 1.1 cont'd

Century	
3rd C.	*Table Talk of Confucius.* (compiled from earlier sources)
	Mr Li's Record of Drugs. (Extant only in quotations)
	Record of the Strange Productions of Lui-hai's Soils and Waters. (natural history)
4th C.	*A Prospect of the Plants and Trees of the Southern Regions*
5th C.	*A Treatise on Bamboos.*
	Collected commentaries on the *Classical Pharmocopoeia of The Heavenly Husbandman.*
6th C.	*Codex Aniciae Julianae* (A re-issue of Dioscorides).
7th C.	*The New Pharmacopoeia.* (preserved in fragments via Japan)
	The Natures of the Vegetable and Other Drugs in the Pharmaceutical Treatises.
8th C.	*Meanings and Pronunciations of Words in Pharmaceutical Natural History.*
	Entry into Learning (encyclopaedia)
	The Manual of Tea
	A supplement for the Pharmaceutical Natural Histories
9th C.	*The Story of Camel-Back Kuó the Fruit Grower.*

Development of Printing

Century	Chinese Works	Western Works
10th C.	Drugs of the Southern Countries Beyond the Seas. Grades of Tree-Peonies of Yueh. Treatise on Bamboo shoots. Master Jih-Huas' Pharmacopoeia.	From Islam - Avicenna The Canon of Medicine.
11th C.	Illustrated Pharmacopoeia. Account of the Tree-Peonies of Loyang.	
12th C.	Imperial Medical Encyclopaedia. Essay on the Tree Peonies of Chheuchow.	
13th C.	Fundamentals of Agriculture and Sericulture. Materia Medica in Mnemonic Verses. Natural History of the Lu Chhan Yen.	Albertus Magnus–De Vegetabilitus Bartholomaeus Anglicus – Liber de proprietatibus rerum
14th C.	Further Advances in Materia Medica. The Beetle and the Sea. (a Biological book)	Development of Printing
		RENAISSANCE
15th C.	Pharmaceutical Natural History of Southern Yunnan Treatise on Wild Food Plants for Use in Emergencies.	Konrad – Das buch der natur
16th C.	Essentials of the Pharmacopoeia Ranked according to Nature and Efficacy The Great Pharmacopoeia Objective Natural History of Materia Medica: a True-to-Life Study.	Brunfels – Herbarum vivae eicones Bock – New Kreutter Buch
17th C.	A Treatise on Edible Wild Plants for Emergency Use.	Grew – The Anatomy of Plants Ray – Methodus Plantarum

them. Treatises dealing with these topics include the *Chiu Huang Pen T'Shao* (1406) together with others published in 1582, 1591, 1622, 1642, 1571 and 1652 which included usefulness in emergencies, even if it were not the main thrust of the book.

In the 20th century, Vavilov (1920) drew attention to the long list of plants consumed in China. It seems not unreasonable to connect this with the Esculentist Movement and it is significant that in a Chinese context it might be possible to elucidate a sophisticated literary basis rather than assume the results of unrecorded trial and error surviving only through folklore. This matter is considered in more detail in Chapter 4. Appendix 1 summarises the contents of the *Chiu Huang Pen TShao* and possible links to Vavilov.

Sufficient perhaps has been said to indicate something of Chinese plant science from the days preceding contact with Europe. One might jump to the conclusion that the richness of Chinese plant life gave the primary impetus. That hardly seems likely, although it must have been a subsidiary factor. The literature considered here should be seen as part of a much wider cultural tradition since, contemporary with these books, were those produced on a wide variety of subjects including astronomy, calendars, calligraphy, dictionaries, education, government, history, mathematics, orthography, philosophy, sport, textual criticism and travel besides many others. It is as if the concentration of intellectual excitement that occurred in Europe at the Renaissance were instead played in slow motion over a far longer time scale in China.

Chinese Plant Illustration

Some Chinese classics are embellished with plant illustrations and a modern author could, if he chose, include reproductions of these. A word of caution is appropriate. The Chinese were adept at reprinting books repeatedly and, depending on the direction of, say, imperial patronage, earlier illustrations might be redrawn or even added to later editions when they were not present in the originals. Modern techniques allow facsimile reproduction but clearly there is value in locating for this purpose surviving editions which are as close to the original as possible, provided, of course, that the illustrations are informative and sufficiently accurate to justify the process. Scholarly awareness of these things nowadays precludes incorporation of "traditional" illustrations merely to veneer the text. For a discussion of Chinese plant illustration see Haudricourt and Metailié (1994) and Metailié (1995)

A First Assessment

The contrast during the Dark Ages of Europe with China is not readily explained. During that time the scholarly tradition was largely the preserve of the Church. Its language was Latin, making access to Roman writers straightforward, and its preoccupation with Greek philosophy might have kept open the door for scientific initiatives begun by Aristotle and his pupil Theophrastus, but, it was not to be. *The Canterbury Tales* Chaucer (an English poet) were published around 1387. Among its stories is that of the Nun's Priests Tale where a cock, Chauntacleer, being poorly, is advised by his wife, the hen Pertelote, of the various remedies she has on offer – all it seems, derived from Dioscorides nearly 1400 years previously. Should a Buddhist pilgrim in the 14th century have chosen, on a visit to one of the shrines in China, to have passed the time with a similar story, he would have been able to refer to a far more comprehensive and up–to–date pharmacopoeia.

The Modern Period

Until about 1550 contact between China and Europe was spasmodic. Westerners, among them Jesuit missionaries and traders, thereafter began to open up the situation. There was no sharp transition but the year 1800 provides a convenient marker. Significantly, it is after then that traders began to recognise and respond to the taste for "chinoiserie" becoming fashionable among Europeans with sufficient disposable income.

Between 1550 and 1800 plants from China began to be introduced into European gardens among them *Althaea rosea* (hollyhock), *Camellia japonica, Gardenia grandiflora, Jasminum nudiflorum* and *Wisteria sinensis*. Part of the stimulus to find hitherto unknown plants came from Linnaeus, although, nowadays, we might challenge his too–ready adoption of "orientalis" or "chinensis" as specific names for things originating east of Suez.

After 1800 the rate of plant introduction increased and during the 19th century *Davidia involucrata* (the dove or handkerchief tree), *Forsythia*, and *Trachycarpus fortunei* (Chusan or windmill palm) were added to our gardens. Something else though, of considerable significance, had begun to happen. This was that, having discovered jasmine, camellia, forsythia and other plants Western botanists and nurserymen started to explore the *range* and *diversity* of these now familiar plants. The Chinese flora did not disappoint. Those famous collectors, Farrer, Forrest, Kingdon-Ward and Wilson among them, enormously enriched our choice of garden plants, a matter considered in more detail in Chapter 4.

Floristic Richness

A recurrent and indeed major theme of this book is the richness of the Chinese flora, and it is appropriate at this point to whet the reader's appetite in a way that provides a helpful orientation. Although any number of plant taxa would do for the purpose, the example chosen here is *Jasminum* already mentioned, in the family Oleaceae using data from the *Flora of China* (Wu and Raven 1996).

On a world basis there are about 200 species of *Jasminum*. Of these, 43, or nearly a quarter, occur in China. Among those 43 are 19 endemics whose natural distribution is confined to China. It is, however, instructive to pursue the matter further and see how, within China, these various species are distributed.

Of the 43 species occurring in China, almost all of them are represented in the south–western province of Yunnan, (35/43). In Sichuan, 12/43 are found. If we next consider just the 19 endemic species, 14 of them while they may occur in various parts of China, nonetheless are found in Yunnan. The endemic *J. nintooides* is, in fact, known only from that province. Closer scrutiny of the list shows that Yunnan and Sichuan monopolise the lion's share not only of the 43 Chinese jasmines but also, more particularly, of its endemics. Staying only within Oleaceae, the World and Chinese figures for species numbers include the following *Fraxinus*, 60:22, *Forsythia* 20:6, *Syringa*, (Lilac) 20:16, *Osmanthus*, 30:23, *Olea*, 40:13, *Myxopyrum*, 4:2, and *Ligustrum* (privet) 45:27. Were we to probe these in detail, the richness of Yunnan and Sichuan would again become apparent. Indeed, if we extended matters to the whole of the supposed 30,000 species of Chinese plants *almost half* would be shown to occur in Yunnan, and with Sichuan to be an area of most remarkable diversity.

In seeking, therefore, to provide a rationale for the Chinese flora we need to show not only how there comes to be such a species diversity but, additionally, how it is so impressively concentrated in particular areas. Is it that mountain regions are congenial to speciation, that elimination has occurred elsewhere or some combination of these? As might be assumed, landscape and climate are highly significant, and to these we now turn.

Note: Since this chapter was written, a further part has been added to Volume 6 of *Science and Civilisation in China*. This is part 5 by Huang H T (2000) and deals with Fermentations and Food Science.

2 Landscape and Climate

Topographically, a simplified view of China would consist of an awareness of such major features as, to the west the Himalayas, in the north the Loess Plateau fringing the Gobi Desert and, as one moved south the Yellow and Yangtse rivers issuing east and separated by the Tsinling Mountains. Finally, one could sweep an arc along the whole length of the China coast noting its hinterland of seaward mountains, plains and deltas.

From this readily grasped starting point one can then add two kinds of detail. The first is merely more topography, which qualifies the original overview. The second, recognising that the present is the outcome of past geomorphogenesis, provides information about the form and content of present–day land masses. Both kinds of detail are necessary but it is essential to retain a sense of perspective.

Because this book is so largely preoccupied with the angiosperms, the flowering plants, geology from the remotest recesses of time in the Earth's formation is hardly relevant, and it is appropriate to begin here with the situation as it was at the dawn of the Cretaceous period. If by this time the angiosperms existed, they were inconspicuous and, in any case, their time of genesis remains elusive. What is more to the point is that during the Cretaceous, the flowering plants achieved much of their astonishing diversity. Continuing into the Tertiary, they did this against the panorama of continental drift and climatic change. Of course continents had been altered before but it is this later part of the continental symphony accompanying the rise of the flowering plants that primarily interests us here.

China and the Timetable of Continental Drift

If the major plant families and their principal subdivisions are present on all continents then their evolution and establishment could be presumed to *pre*–date the breakup of Pangaea. If, conversely, the lower levels of the hierarchy are more confined, then their origins and subsequent history presumably *post*–date continental separation and modification. As Clayton and Renvoize (1986) comment:

"Evidently, the genera are not good travellers..."

Until, that is, that most mobile of creatures, humankind, began consciously or otherwise, moving them around. Primarily, therefore, to grasp the impact of land formation on the Chinese flora the principal events are shown in Table 2.1.

Prior to the collision of India with Eurasia some 40 or so million years ago, it is important to recognise the existence of the Tethys Sea reaching across much of the Old World. Its diminutive surviving remnants are the Mediterranean, Black, Caspian and Aral Seas. Tectonic changes by such events as the northward movement of India drove the Tethys Sea floor beneath what was to become the Himalayas.

Table 2.1. Major post-Jurassic events relating to China

Period	Epoch	Duration MY	MYA	Event
	Holocene	Approx. last 10000 years		← Rising sea level re-isolates Taiwan and later, Hainan
Quaternary		Approx. 25000 years ago		← Lowering sea level joins Hainan and, later, Taiwan to the mainland
	Pleistocene		2.5 ←	Himalayas now earth's highest mountains – glacial phases especially in Europe and N. America
			2.5 ←	Origin of the loess plateau accumulation
	Pliocene		4.5 ←	Climatic downturn
			7	
	Miocene		19	
			26	← Renewed activity in Kunlun / Tsinling, Altai and Tienshen mountain systems
Tertiary	Oligocene	12		← Significant raising of Himalayas
			38	
	Eocene	16		← Collision of India with Eurasia
			54	
	Paleocene	11		
			65	
				← Origin of the Gobi desert
Cretaceous				Origin(?) and substantial diversification of Angiosperms.

The arrival of India continued the underthrusting but, there was added the battering–ram effect of the subcontinent. The result eventually was to create not only the Himalayan ranges but, too, the Tibetan plateau. Curiously, marine deposits from the Tethys, because of back-folding of strata can be found among the Himalayan peaks.

The geomorphic processes which led to this situation continue and it is only within the last 600000 years that the Himalayas have become the world's highest mountains.

The Gobi is an old desert, having existed for more than 65 million years; its beginnings coincide with the demise of the dinosaurs, and their fossilised remains, though localised, are relatively abundant there. With the rise of the Himalayas, a strengthening of the monsoon winds led to the formation of the Loess Plateau stretching across northern China. This is wind–blown dust from the desert accumulated over about 2½ million years to a depth of up to 100 m. For an account of the formation and evolution of the Loess Plateau see, for example, Zhang and Dai (1989). Presently, material is being both added and subtracted, the material removed in large measure finding its way into the Hwang Ho, the so–called Yellow River – hence the name. This river, rising in the Pa-yen-k'a-la Mountains of Tibet, on its journey through the Loess Plateau becomes the world's muddiest river. It has on average 34 kg of solids of water and deposits into the sea 1.52 billion tons of silt per year. Wang et al (1988) provide estimates for both the deposition and removal of loess at current rates. (For the origin and significance of loess for Korea and Japan see Mizota et al (1991).

Moving south to the Tsinling Mountains, these, like the Kunlun range, of which they are an eastward extension, consist of crystalline rocks some 250 million years old and separate the loess from the warmer, greener south of China through which the Yangtze River flows. The Tsingling Mountains together with the Kunlun, the Altai (in Inner Mongolia) and the Tien Shan (running southwest–northeast along the border with Kyrgistan) all underwent a new phase of activity from about 26 million years ago.

At 6300 km, the Yangtze is the world's third longest river. Curiously it has two river heads, both in Tibet, at Tang-ku-la Mountains and Wu-lan-mu-lun and two exits to the sea, dividing either side of what is now Ch'ung Ming Island created about 1000 years ago. The river goes through Sichuan and eventually the three famous, difficult–to–navigate gorges, now the subject of a massive water control scheme. This is in part necessary due to the propensity of flooding in a river basin which provides 70% of China's rice in an area of high population density.

Both of these major rivers, the first conspicuously so, in depositing silt, create both mobile deltas and, as a result, alter, through time, the coastline.

Taking a longer time perspective, the coastline of China has changed in response to altering sea levels as a result of glaciation mostly elsewhere. At the extremes of glacial advance, world sea level fell appreciably by about 180 m some 237000 years ago. For China the situation can be summarised as follows. At about this time Taiwan and Hainan were linked to China. With rising sea levels Taiwan became re-isolated by about 10000 B.P. as did Hainan subsequently about 6000 B.P. The consequences for animal migration of climatic change are examined by Huang and Chen (1988), Xu (1988) and Zhang (1988).

Seismic Activity

Although there are no active volcanoes in China, given the immense and recent geological uplift of the Himalayas and the Tibetan Plateau it is hardly surprising that western China shows substantial seismic activity. To the east the "ring of fire", that line of active volcanoes around the Pacific Rim, does not reach the Chinese mainland, although two Philippine volcanoes are within about 500 km. More significantly, the converging boundaries of the Philippine and the Eurasian plates render Taiwan seismically very active.

Some Chinese earthquakes are severe. The mining and industrial city of T'angshan about 110 km east of Beijing was almost destroyed in the earthquake of July 28, 1976. More than 240000 people died and about 500000 were injured. More recently, there have been other earthquakes including, for example, in 1995 Yunnan 6.5, 1996 Yunnan 7.00, Xinjiang 6.90, Inner Mongolia 5.90 and in 1997 Xinjiang 6.40. In 1999, there was a severe earthquake in Taiwan in which 2400 people died.

Not only seismic activity disturbs the Chinese landscape. There are, too, repeated instances of severe weather, and these will be considered shortly.

A Botanical Preliminary

One aim of this book is to identify themes, that is to bring to the reader's attention underlying aspects which help explain the rich diversity and geographical distribution of the Chinese flora. Yunnan is, botanically, spectacularly rich, and the preceding discussion on the formation of the landscape provides a highly relevant way of introducing that province as an example. While many writers have described various aspects of plant diversity in southwest China, few have sought to grapple with the texture of flowering plant evolution there in so stimulating and fundamental way as Takhtajan (1969). Among the features he identifies, the following particularly are worth recalling. Of one area of forest he studied he remarked that the floristic composition was exceedingly diverse with the co-existence of many tropical and Holarctic genera. He noted in particular "the abundance of primitive forms as to both the angiosperms as a whole and individual orders and families". Of Yunnan and the adjoining areas (eastern Himalayas, Assam, upper Burma, North Vietnam and south Japan), he remarks that the flora includes "almost all the basic phylogenetic groups which provided material that served as the basis of the temperate floras of the Northern Hemisphere". He quotes Federov in regarding the forests of Yunnan as "clearly transitional being intermediate in character between tropical and Holarctic vegetation".

Using this as a starting point we stress two facts. The first is that, for whatever reason, ahead of Tertiary mountain building the area was already botanically rich. Secondly, with the results of such mountain building to hand, and with it the proliferation of myriad diverse microhabitats, a situation was created sufficiently benign to permit both the survival of primitive forms and provide niches congenial to newly evolving variants. In summary, Yunnan, already botanically rich, was made richer by the results of its tumultuous geological history. A fragmented landscape has provided there a garden of evolution.

Is it possible to examine this situation at some intermediate point in the sequence of events over many millions of years? Consider the following example from an Oligocene deposit in Yunnan at the Jing'gu Basin. Macrofossils of *Annona*, *Uvaria* (Annonaceae), *Machilus*, *Nothaphoebe*, *Phoebe* (Lauraceae), *Zelkova* (Ulmaceae), *Ficus* (Moraceae), *Lithocarpus*, *Quercus* (Fagaceae), *Myrica* (Myricaceae), *Carya* (Juglandaceae), *Rehderodendron* (Styraceae), *Rosa*, *Sorbus* (Rosaceae), *Cercis*, *Erythrophloeum* (Leguminosae), *Terminalia* (Combretaceae), *Rhus* (Anacardiaceae), *Oreopanax* (Araliaceae) and *Jasminium* (Oleaceae) were found. Such a flora would be mostly evergreen broad leaved but mixed with some deciduous species resembling that presently found in parts of tropical south east Asia (Li and Zheng 1995). Those genera marked* are represented in the present–day flora of Xishuangbanna in southern Yunnan, for example, (Li 1996).

A changing landscape has had botanical consequences and that same landscape has locked in its fossil record clues to the succession of floras which have clothed it. Subsequently, we return to this topic at various points.

Climate

Geological change is intimately involved with climate change. The present climate is, in China as elsewhere, merely the currently prevailing one in a succession, a flux, from the past and stretching into the future.

Any plant species has a range of adaptation and if we find its fossil relatives we are, with due reservation, able to make assumptions about the corresponding palaeoclimate in which they grew. If, subsequently, they are attenuated in the fossil record and replaced by plants with different environmental requirements, again, we can infer a palaeoclimatic change. For China, an accessible treatment utilising this approach is that of Zhao (1992) from which the section which follows is substantially derived. It depends on two types of fossil – macro, deriving from plant organs and micro, primarily spores and pollen grains. These have to be carefully related to the geological context within which they occur.

It is necessary to point out some major considerations. The Chinese land mass changing as it did, is inseparable from the global situation. Again, so large a land mass is involved that contrasted regions could coexist as, of course, they still do. It needs, too, to be noted that fossil sequences are normally discontinuous and that our attempts to tell a continuous story are based on fragmentary evidence. Finally, whether world temperatures were rising or falling, the rise of the Himalayas and the Tibetan Plateau provides an increasingly influential aspect through time on Chinese climatic evolution.

A Palaeoclimatic Digest

1. Cretaceous

Given the concentration of this book on angiosperms, as with geology, we begin with Cretaceous. At this time the Tethys Sea separated the Indian and African land masses from Eurasia, allowing the circulation of the equatorial circular current and this at a time of high temperature and uniform climate. A relatively rich flora reached Alaska, Greenland, Spitsbergen and Siberia, for example. What subsequently became the Tibetan massif was at or below sea level and attached to the Indian land mass.

Moving through the later stages of the Cretaceous, the tendency was for the climate to become drier. Eventually, too, the gymnosperm/pteridosperm vegetation began to acquire an appreciable angiosperm content. In south China, for example, pollen of *Acer, Fagus* and *Juglans* became detectable, indicating here the interpolation of a wetter phase.

2. Tertiary

With the dawn of the Tertiary the Indian plate moved with increasing speed, eventually closing the Tethys Sea and colliding with Eurasia. During the Palaeocene the temperature, globally, rose perhaps a further 9°C higher than at present. Thereafter it declined toward the Quaternary.

During the Tertiary, the eastern coast of China moved further eastward several hundred kilometers. The low-lying regions Tibet and Yunnan would have had an approximately mediterranean–type climate. Later, in the Eocene, Yunnan contained such evergreen genera as *Cinnamomum, Dryophyllum, Ocotea* and many others, with a further enrichment of the flora during the Oligocene.

With the arrival of the Neogene (Miocene plus Pliocene), the rising land mass to the southwest and west was increasingly contributing to climatic contrast within China. The eastern temperate region acquired deciduous genera such as *Betula, Corylus, Liriodendron* and *Ulmus*, indicating strong seasonality. By this time, upland species of *Quercus*, together with *Trapa*, were detectable in the Tibet region together with *Abies* and *Picea*.

3. Quaternary

It was during this period that the Ice Ages occurred, devastating the floras of the higher latitudes of both the Northern and Southern hemispheres. Global temperature was 5-7°C lower than at present. The Qinghai-Tibet Plateau continued to rise, stimulating the east Asia monsoon and exerting significant influence now on the climate of southeast Asia. As temperature declined during the Pleistocene, so glacier formation developed but not on a scale comparable to elsewhere in the world. Contemporary within this there were to be found in Yunnan, for example, representatives of Palmae, Moraceae, Myrtaceae and Santalaceae, indicating a substantially tropical climate. During the early Pleistocene (2220–1200 Ka B.P.) a warmer phase intervened and the developing Loess Plateau in N.W. China, for example, became wetter. From 1200–730 Ka B.P. a cooler phase occurred. Between 730 and 130 Ka B.P. two warm and two cold phases followed and then between 130 and 11 Ka B.P. there was the driest and coldest stage in the Quaternary. The Qinghai-Tibet plateau had risen to 3000 m above sea level, influencing atmospheric flow across China.

4. Holocene

In China, as elsewhere, the average temperature rose and paralleled changes taking place in other parts of the world namely cool temperate, warm and wet, dry and cool. Around 7000 years ago occurred the so-called Atlantic phase a benign climatic interval with temperatures 3–4 °C higher than at present and with higher rainfall. It was around this time that Chinese culture began to take shape. Later, the so called Little Ice Age familiar to Western historians had its counterpart in China from 1550 - 1850 reaching its extreme in the latter half of the 17th century. For an account of this see Wang (1992).

Since about 1850, with the increasing influence of the Industrial Revolution, levels of atmospheric CO_2 have risen, tending now to detectable global warming, a situation from which China is not exempted. Lastly, if the ozone level in the upper atmosphere continues to be diminished, China as, elsewhere, will detect and, be adversely affected by, increasing ultraviolet radiation.

As was pointed out earlier, it is from plant fossil distribution that we partly derive our perceptions of palaeoclimates. Later in this book some emphasis is given to palaeobotany, which goes beyond its significance of discerning climatic change, and concentrates upon the plant interest.

Present Day Chinese Climatic Diversity

The outcome of geomorphological change across China is present–day climatic diversity. The situation can be summarised as follows.

Table 2. 2. Climatic parameters for ten Cities

City	Province	Lat N/ long E	Altitude (m)	Climate
Urumqui	Xijiang	43°40'/87°41'	913	Semi desert. Hot summers, cold winters Sparse rainfall spread through the year
Lanzhou	Gansu	36°02'/103°42'	1508	Continental Hot summers, cold winters Sparse rainfall spread through the year
Xian	Shaanxi	34°16'/108°55'	412	Continental Hot summers, mild winters Moderate rainfall and least at the turn of the year

Table 2.2 cont'd

Beijing	Beijing	$39^0 56'/116^0 20'$	52	Continental Hot summers, mild winters Moderate rainfall and concentrated around June, July especially
Harbin	Heilongjiang	$45^0 41'/127^0 08'$	143	Continental margin. Hot summers, cold winters. Moderate rainfall and concentrated in the middle third of the year
Lhasa	Xizang	$29^0 33'/91^0 04'$	3685	Dry, cold, montane Cool summers, cold winters. Moderate rainfall and tending to fall in the middle six months of the year
Kunming	Yunnan	$25^0 05'/102^0 43'$	1893	Sub tropical Warm summers, mild winters Wet most months except at the turn of the year
Chungking	Sichuan	$29^0 23'/106^0 29'$	261	Subtropical Hot summers, mild winters Wet most months except at the turn of the year
Shanghai	(Auton.)	$31^0 18'/121^0 29'$	4	Subtropical Warm summers, mild winters Wet throughout the year but more so in summer
Hong Kong	(Auton.)	$22^0 23'/114^0 90'$	0–958	Tropical monsoon Warm summers, warm winters Very wet most months

From north west to south east, rainfall increases and from north to south, winter temperatures rise. It is therefore possible at a fairly low level of resolution to show that across China there are concentric areas of increasing plant productivity as shown in Plate 1.

At a higher level of resolution within this general pattern one can then select ten cities to indicate something of their climatic individuality such as is shown in Table 2.2.

The modulation of temperature throughout the year depends largely upon proximity to the centre or the margin of a continental land mass. Thus, Urumqui and Lanzhou, for example show a wide amplitude between winter and summer (from below minus 10° to 30°C plus respectively). At the opposite extreme the situation is evened out in Hong Kong (from 12°C to 30°C correspondingly). In terms of rainfall, Hainan and Hong Kong belong to the moist tropics but do not have the heavy year–round rainfall of the wettest tropical regions of the world.

Winds can be either dry or wet, depending on whether they have traversed land or sea and across China, as elsewhere, wind patterns vary throughout the year. In winter, drying monsoon winds bear down on China from central Asia. In summer, moisture-laden monsoon winds blow across the sea from the southeast, while to the north, westerlies blow from central Asia. To a large extent this explains the wetter summers and drier winters so characteristic of much of China. It is necessary to take account of the North Equatorial Current, which, since it comes from the south, moving northward, warms the coast, and, to some extent, the southeastern hinterland. It provides an effect approximating to that branch of the Gulf Stream, the North Atlantic Current moderating the climate of the British Isles which otherwise for its latitude would be more typically northerly.

The pattern set out here can be modified by altitude the extremes being, (Table 2.2), Shanghai at sea level and Lhasa at more than 3500 metres. The effect on plant life is particularly marked in Yunnan, with low-lying tropical forest to the south and, at its northern end, alpine plants growing up to the snowline.

From Climate to Weather

Whatever may be discernible through time and expressed (as in Table 2.1), this gives only a partial, though nonetheless useful, impression. Essential to our understanding is an awareness of what nature can do in the short term – perhaps over no more than a day or two through driving rain and ferocious winds. One should not draw the conclusion, of course, that such things happen only in China but it is nonetheless useful to be made aware of the scale of calamity which can occur.

Not all of these storms would have reached the Western world's headlines, and after them, the pattern is that the Chinese, with their customary resilience, set about reassembling home and livelihood. One needs to recognise, though, the size of a storm that can stretch its effects across nine provinces, for example.

If one's perception of China were a mist-laden landscape where mandarins stand serenely among the timeless beauty of tree peony and flowering cherry against a background of pagodas and pine trees then, maybe, the awesome statistics of Table 2.3 will induce a note of realism. From such a basis as is provided by these first two chapters we can now turn to the matter of native plant distribution.

Table 2 3. Some incidents in China in recent years

1993	Torrential rains in Hunan and Sichuan causing massive flooding and landslides with about 120 fatalities
1994	Torrential summer rains and flooding in Guangdong and Guangxi, damaging housing industry and agriculture with about 400 fatalities. Later, in Taiwan a typhoon with windspeeds of 137 km/h^{-1} severed power lines and blew down hundreds of trees with 10 fatalities
	In Beijing a suffocating heat wave claimed the lives of 104 people
	In late August in Zheijiang typhoon Fred resulted in severe damage and 1000 fatalities
1995	In Hunan, Hubei and Jiangxi severe rain produced massive flooding destroying 9000000 homes and with 1200 fatalities.
1996	In Yunnan, in 4 days' rains caused two landslides on Laojin Mountain with 100 fatalities

Table 2.3 (cont'd)

1996 (cont'd)	Later, heavy rain across nine provinces caused flooding of the Yangtze and Yellow and Hai Rivers with severe damage to crops and property and with about 2000 fatalities
	In August typhoon Herb caused $507m in damage to agriculture and fisheries and with 41 fatalities
	In September in Guangdong, typhoon Sally smashed 200000 homes with at least 139 fatalities
1997	In Guangdong floods inundated 177 villages with at least 110 fatalities
	Typhoon Winnie created winds of 148 km/h^{-1}, produced heavy rain with 37 fatalities in Taiwan and 140 in Zhejiang and Jiangsu

3 Native Plant Distribution

Far more than one might at first suppose, the subject of plant distribution is one of considerable significance. To see the matter as a mere compilation of what grows where is to miss the point almost entirely. Where, then, do the interest and importance lie?

Perhaps the most convenient point with which to begin is the *genus*. Put colloquially, each genus can have its own way of doing things – where it will thrive best and how it responds to change. Does it have a few "all purpose" species which grow widely distributed or a vast number of species each precisely adapted to its own small specialised habitat? Is a genus in an evolutionary expansionist phase or is it in decline, hanging on here and there but, through loss of habitat or challenge from more aggressive competitors, doomed to extinction? The world's plant diversity is far from evenly spread. If one considers China particularly, how is it that Sichuan and, more especially Yunnan, teem with species compared to elsewhere?

Awareness of diversity can work in more than one way. What engrosses a botanist about a particular genus can guide a horticulturist in the search for new ornamentals or a breeder looking for agronomically useful characters. Conversely, horticulturists combing the world can alert a botanist to the significance of a given region in understanding the diversity of some genus or other.

A Theme for China?

Could one, for China, identify a simple botanical theme and then, region by region, show how it has been orchestrated? Or do we, rather fatalistically, assume that so huge a flora, even if it is the outcome of causes, is too complex for analysis? The former view is taken here, even if it turns out in practice that the effect of human occupation and exploitation is rather like losing parts of a musical score.

The Theme

In early Tertiary times China, across its relatively even surface, was host to a smooth gradation of plant species – tropical to the south, warm–temperate in the north. Subsequent mountain building especially, but not only, of the Himalayas led to the creation of a vast array of microhabitats providing opportunities for those genera with the genetic versatility to respond. The mountain uplift with its modifying effect upon the monsoon resulted in the accumulation of the Loess Plateau, which through dehydration and deposition caused the replacement of a mesophytic flora by one more xerophytic. Since, in China, the glacial effects of the Ice Ages were far less severe than in Western Europe and North America, mere survival of plants in China relative to their elimination elsewhere has been a significant factor. As for other regions, glaciation resulting in lowered sea level has linked Taiwan and Hainan to the mainland for shorter and longer periods, respectively, with detectable consequences for plant migration.

In summary, the theme is one of progressive elaboration in relation to continuing diversification of the region until the rise of *Homo sapiens*, whose effects for the most part have been disruptive of nature's vegetational pattern. Plates 2 and 3 show instances of contrasted Chinese landscapes.

Reservations and Qualifications?

While this simply stated theme might be a useful starting point, several questions soon present themselves. They include the following:

1. Does the available fossil evidence suggest that a sufficiently diverse "starter kit" of vegetation was in place ahead of significant mountain uplift?
2. To what extent might diversity be explained by subsequent migration *into* the region?
3. Is it possible to detect any correlation from the fossil record between increasing plant and increasing habitat diversity through time?
4. Could the pattern of plant diversity across China have more than several causes?

It is not proposed to attempt to provide the reader with definitive answers but rather to point out that these questions have helped influence the choice of examples throughout the book. Nonetheless it is necessary to offer some preliminary comments.

The Fossil Record

There is available in English *Fossil Floras of China Through the Geological Ages* edited by Li (1995) – a compendium of 695 pages together with 144 plates and a substantial bibliography.

Beginning, for our purpose, with the Cretaceous, the following points are significant. The early Cretaceous was dominated by gymnosperms among the seed plants and intrusion of an angiosperm element was minimal. Among perhaps the earliest angiosperms to be detected in the world were those found at Jixi, Heilongjian, in the genera *Asiatifolium, Jixia, Shenkuoa* and *Xingxueina*. All are now extinct.

By the late Cretaceous there emerges from fossil data probably sufficient evidence to support the idea of a distinctive northern and southern flora. Although some genera have not survived, even at this early stage angiosperm genera were found whose descendents have persisted into modern times notably, *Platanus, Quercus,* and *Trapa* in the north, and in the south, for example, *Cinnamomum* and *Nectandra*. Thereafter, more and more surviving genera become apparent in the fossil record. In Xizang, late Cretaceous fossil genera include *Aralia, Ficus, Populus* and *Viburnum*.

The early Tertiary flora of China is considered to have diversified rapidly. Li and Zheng (1995) point out that Himalayan uplift began in the Oligocene and that, while many genera from the Palaeocene and Eocene have become extinct, many from the Oligocene have survived. For Yunnan, at this time, there was a flora with Fagaceae most abundant with ten species, and Lauraceae next with four. Through the Miocene, the Himalayan uplift continued, causing, through its various subsidiary effects, a diversification across China of both topography and climate – a situation reflected in the diversification of the Chinese flora.

An issue inherent in this subject is how far throughout succeeding epochs one can sustain a view of regionalisation of the flora. Floras both preserved in small regions and discovered by chance need not be truly representative of wider trends. Again, if we find, say, an incontrovertible remnant of *Quercus*, what could be inferred about climate? At the present day, for example, hundreds of oak species occur and can be found from Malaysia to Scandinavia.

Moving from the Pliocene into the Pleistocene one major theme is the contrast in susceptibility to glaciation between on the one hand Europe and North America and China on the other.

As the fossil flora grades into that of the present day it becomes evident after Europe emerged from its Ice Ages why 'restocking' it with genera that had survived in China is so evidently a feature of modern horticulture.

A Paradox?

Takhtajan's (1969) comments on Yunnan were indicated in the previous chapter. We need to recall, however, that its present topography is, in geological terms, quite recent. One example will serve here. *Litsea* species (Lauraceae) are known in the fossil record of southwest China from Eocene and Oligocene times (Li and Zheng 1995) and comprise one of the primitive taxa Takhtajan (1969) had in mind. The implication, perhaps, is that interesting primitives were already located here ahead of the Himalayan uplift and consequent topographical diversification. Is it this latter which has here and there provided the benign environment for survival? Paradoxically, is it, too, the case that such a topography is also congenial to further plant innovation and that diversity thus has more than one cause here?

Could we thus explain much of the diversification of *Primula* and *Rhododendron*, selected later for comment?

Endemism

Strictly, an endemic species, genus or family is one confined to a given area. On this basis it is possible to find taxa which are confined to China. In one sense this can be misleading, since China is a politically defined area quite independent of any botanical considerations. If therefore a plant is restricted to, say, Sikkim and the adjoining part of China, both together comprising a relatively small area, what is the endemic status of that plant? Botanically, it might be more realistic to define an area called Sino-Japan spilling over into the eastern Himalayas, and to ask whether or not plants were confined within that. With that reservation, of what significance is endemism within the Chinese flora? A recent compendium is *The Endemic Genera of Seed Plants of China* by Ying et al published in 1993. The authors distinguish paleoendemics, taxonomically isolated and found in isolated refugia; schizoendemics, derived through graded speciation from a common ancestor; patroendemics, which are parental diploids that have given rise to polyploids; apoendemics, polyploids having arisen from widely distributed diploids and lastly, neoendemics, recently derived and not having spread beyond their region of origin.

Clearly, once endemics have been recognised, together with the likely subcategory to which they belong, a facet of the innate character of a regional flora begins to emerge. Is a given taxon a 'living primitive' endemic by reason of hanging on in a benign environment or is it diversification prompted by a change of circumstances coupled with the ability to respond?

Not only by reason of its large land surface but because parts of the Chinese flora are considerably diverse, there is a large number of endemic genera and even two or three endemic families. (Endemic species run to many hundreds).

So as to build up for the reader some awareness of endemism in the Chinese flora, considerable attention is directed to this in various parts of the book when dealing with particular groups.

A Sense of Locality

Whether large or small, in order to introduce the plant life of any area it is necessary to generalise; but for any individual making contact with that flora, his or her awareness necessarily will be partial. No one locality is quite like another and only with considerable experience can one sort out the incidental from the significant. Contrast can help and there follow now some cameos of particular parts of China together with some indication of the questions they prompt.

Although Yunnan is rich in species it is far from uniform and here attention focuses on Xishuangbanna in its southwest.

Xishuangbana, Yunnan

This region in southwestern Yunnan against the border with Burma (Myanmar) is botanically a very rich part of China. It is essentially tropical forest well supplied with diverse taxa by nature to which the activities of humankind have added others. A problem in describing its floristic richness is how to convey what is remarkable without burdening the reader with literally dozens of unfamiliar family names let alone those of hundreds of genera. Table 3.1 therefore provides the expedient of grouping families according to the number of genera they have present in Xishuangbanna. The first column groups those families having from one to five genera present in the region, column two those with six to ten and so on.

Simple inspection of such a table indicates that the large majority of families are represented by five or less genera, in reality usually one or two. It is evident that not only angiosperms but dicotyledon types dominate the situation numerically.

Having established this relatively straightforward assessment it then becomes more manageable to examine a few aspects in more detail.

Among the pteridophytes, the two families with more than five genera present are Aspidiaceae (10) and Polypodiaceae (14). Since so few gymnosperms present it is practicable to name each family – Araucariaceae (1), Cephalotaxaceae (1), Cycadaceae (5), Ginkgoaceae (1), Gnetaceae (1), Pinaceae (3), Podocarpaceae (1), Cupressaceae (8) and Taxodiaceae (6).

Angiosperms

It is among these that there is a welter of complexity and the aim must be to highlight the principal considerations.

As might be expected, many of the families represented by just one or two genera are those primarily temperate or warm–temperate near the southern limit of their range, such as Betulaceae (2), Caprifoliaceae (2), Hypericaceae (1) and

Primulaceae (2). Others are simply small families with few genera in total, of which obvious examples are Balanophoraceae (1) and Nepenthaceae (1). Again, there is the influence of differing taxonomic opinion and, for example, whether Lobeliaceae (1) is recognised or merged with Campanulaceae inflating that from 8 to 9 genera found in Xishuangbanna.

Table 3. 1. Genera represented within Families for Xishuangbanna (from Li 1996)

Genera/Family		1-5	6-10	11-15	16-20	21+	
Pterodophytes		39	2				41
Gymnosperms		7	2				9
Angiosperms	Dicots	158	21	13	8	8	208
	Monocots	26	2	3	1	4	36
	Totals	230	27	16	9	12	294

Where families with more genera present are studied, as might be expected, some are more obviously tropical, such as Icacinaceae (7) and Melastomataceae (8), and so it is something of a surprise to find such as Fagaceae (6) and Ulmaceae (6). However, as numbers of genera rise higher for particular families, less surprisingly one finds the more obviously tropical families well represented such as Annonaceae (17) and Sterculiaceae (18).

As regards the monocotyledons, the numbers of families in the various categories are smaller and trends less discernible. Across the whole range are families familiar to a botanist used to a temperate flora and, if one takes into account the efforts of horticulturists, representations of several others that we recognise as garden subjects.

Reference was made earlier to human activity augmenting the flora of Xishuangbanna; there are obvious examples. The Bromeliaceae and Cactaceae, for instance are, in nature, entirely New World families. Similarly, the Poaceae for example is inflated by the presence of *Zea* here, as in virtually every sufficiently warm region.

The More Numerically Represented Families

Among the angiosperms, particularly, it is instructive to consider the most conspicuous families numerically. These are the following and include those of 20 and more genera. The numbers are presented against their approximate world total of genera in Table 3.2.

It is evident that each family is, on a world basis, well supplied with genera. All the families except perhaps the palms could be described as cosmopolitan and here as elsewhere the adaptive combination of their floral structures testifies to the consummate success of the four emboldened families.

Against the background of Xishuangbanna one can then consider other contrasted regions of China.

Table 3. 2. The more numerically conspicuous families at Xishuangbanna

Dicotyledons		Monocotyledons	
Acanthaceae	32/250	Araceae	24/115
Apocynaceae	33/180	Liliaceae	20/250
Asclepiadaceae	25/130	**Orchidaceae**	91/735
Compositae	64/900	Palmae	36/217
Euphorbiaceae	43/300	**Poaceae**	73/775
Labiatae	31/180		
Leguminosae	94/600		
Rubiaceae	42/500		

The Limestone Forests of China

China possesses the largest limestone area in the world, with pure carbonate substrate covering an area of 283000 km^2, mainly ranging from SE Yunnan and Guangxi to S Guizhou (Xu 1986, 1993a). Beautiful karst landscapes and numerous fantastic caves provide various ecological niches for plants and wildlife, as well as for tourism. The limestone forests in China are well known in their rich and very diverse flora, especially the richest in endemism ever recorded in the world. The limestone flora is one of the most important parts of the flora of China and remarkably distinctive from that of others in China. According to the checklist of limestone species (Xu 1993b, 1995), 4287 species and infraspecies taxa of vascular plants have been recorded from the limestone forests in southern and southwestern China. They belong to 1213 genera and 195 families, (Xu 1993b). The limestone flora possesses a higher percentage of various tropical elements – 69.1% of the total entire flora. These tropical elements include pan-Tropical, disjunct in Tropical Asia and tropical America, Old World tropical, Tropical Asia to Australia, Tropical Asia to Tropical Africa, and Tropical Asia. Although the limestone area is within the tropical and subtropical zones, its flora includes 19.8% northern–temperate and East Asian elements. Clearly, the limestone area is rich in plant diversity. Many may be relict species, and some of them are endangered, (Xu and Sun 1984, Xu 1993a, 1995). Three relict Cycads – *Cycas baguanheensis, C, panzhihuanensis* and *C. micholitzii* were, respectively, reported from southwestern Sichuan and southwestern Guangxi. *Amentotaxus argotaenia* var. *brevifolia* is a lonely relict situated at the top of a limestone hill in southern Guizhou. Only two specimens of *Lysimachia scapiflora* were collected from southwestern Guangxi, (Chen 1986), (Liang et al 1981, 1985). More relict species have survived in the limestone forests than in the adjacent acid soil forests. Twenty-two monotypic or oligotypic genera have been recorded as endemic to the limestone flora and most of them are confined to narrow areas, such as *Calcareoboea, Tengia* (Gesneriaceae), *Excentrodendron* (Tiliaceae), *Malania* (Olacaceae) and *Parepigynum* (Apocynaceae). In the limestone forests, about one-third of the species are endemic or subendemic.

Wuyi Mountain Nature Reserve, NW Fujian and SE Jiangxi

The Wuyi Mountain Nature Reserve is located on the boundary line between Fujian and Jiangxi province (NW Fujian and SE Jiangxi), 27°35' – 27° N by 117° 24' - 117° 51' E. It is the highest mountain in SE China – 2158 m. above sea level. The Wuyi Mountain Nature Reserve, an MAB Biosphere Reserve, is divided into a core zone of about 350 ha and an experimental zone, and has a rugged topography comprised of towering peaks and rift valleys. Mt. Huanggang, the highest peak of the Wuyi Mountains, forms a natural climatic barrier blocking off the cold air masses from the north in winter and trapping warm, humid coastal air in summer. Natural vegetation is dominated by large areas of well preserved subtropical broadleaf evergreen forests with a coverage of 95%. The reserve is the southern limit of *Aster tataricus, Anaphalis* spp., *Achillea* spp. and other members of Compositae, and is a transitional zone for flora of the Holarctic and Palaeotropical regimes. There are about 191 families, 755 genera and over 1800 species of vascular plants in the Wuyi Mountains Because it is located along the coast of the E. Sea of China with warm and humid climatic conditions, the total number of species is more than those of any other region at the same latitude, (Lin et al. 1981a, b).

Table 3. 3. Comparison of different sizes of genera and families of vascular plants in the Wuyi Mountain

Number of species	Genera		Families	
	No.	%	No.	%
31-		0.00	13	6.81
21 – 30	3	0.40	14	7.33
10 – 20	18	2.38	27	14.14
5 – 9	64	8.48	30	15.71
2 – 4	234	30.99	54	28.27
1	436	57.75	53	27.75
Total	755	100.0	191	100.0

The genera and families containing one or two–four species are respectively 88.74 and 56.02% in the vascular plants. Presumably, the high percentage of rare and single species in genera and families demonstrates a relict characteristic of the Wuyi flora. The famous relict species *Bretschneidera sinensis, Chimonanthus praecox* and *Camptotheca acuminata* have been recorded in the Wuyi Mountains. The forest species are mainly in Fagaceae, Theaceae, Magnoliaceae,

Lauraceae, Graminae, Rosaceae, Compositae, Leguminosae, Labiatae, Rubiaceae, Cyperaceae, Liliaceae and Euphorbiaceae , (Lin et al. 1981a, 1981b).

Hengduan Mountains Southeast to the Tibetan Plateau

The Hengduan Mountains are also world famous in richness of flora. They are located in the area southeast to the Tibetan Plateau, ranging from SW Gansu, SE Qinghai through W Sichuan to E SE Xizang and NW Yunnan, (Wu 1988; Li and Li 1993 and Zhang 1998). In this region, mountains are generally 4000 m high, and many peaks exceed 5000–6000 m, while neighbouring river valleys are dissected deeply and, thus, relative height differences may reach 2000 m, thereby forming a landscape of high mountains alternating with deep valleys, (Wu 1998;, Li and Li 1993 and Zhang 1998). This situation, combined with the influences of the Pacific Southeast Monsoon and the Indian Southwest Monsoon, makes this region complex with abundant and special vegetation types. On the south slope of the peak of Namjagbarwa (7782 m) where the Himalaya and Jengduan join, within a short distance, are distributed lower mountain evergreen monsoon rainforests dominated by *Dipterocarpus turbinatus* and *D. macrocarpa* (lower than 600 m), lower mountain semi evergreen monsoon rainforests dominated by *Terminalia myriocarpa* and *Altingia excelsa* (600–1000 m), middle mountain evergreen broad-leaved forests dominated by *Cyclobalanopsis lamellose* and *C. xizangensis* (1800–2400 m), middle mountain evergreen coniferous forests dominated by *Tsuga dumosa* (2400–2800 m), subalpine evergreen coniferous forests with *Abies delavayi* and its variety *A. d.* var. *motuensis* as the main constituents (2800–4000 m) as well as alpinc shrubland meadows consisting of evergreen *Rhododendron* shrublands and meadows (4000–4400 m), until an alpine periglacial zone is reached consisting of lichens mosses, and a few plants of Compositae, Cruciferae, Saxifragaceae and other families (4400–4,800 m), (Wu 1998; and Zhang 1998).

In the Hengduan Mountains, there are about 226 families, 1325 genera and 7954 species. Generally, the flora of its mountain range is predominantly temperate in nature, (Wu 1998; Li and Li 1993; and Wang and Wu 1993, 1994). In addition, the Hengduan Mountains are very natural region floristically and abundant in species, genera and families, and complex in geographical elements, especially striking in endemism and vicariance. This region is very complex and biologically enormously rich, as it is cut by deep river gorges that have proved barriers to species movements and resulted in high levels of local endemism. Seventy two genera endemic to China are distributed in the Hengduan Mountains – 28% of the total endemic genera of China. Besides, there are 212 general endemic to these mountains, (Wu 1988 and Li and Li 1993). Some of them are newly recognised endemics, such as *Anemoclema* and *Metanemone* (Ranunculaceae), *Dipoma* (Cruciferae), *Diplazoptilon* and *Formania* (Compositae), *Iso-*

metrum (Gesneriaceae), *Skapanthus* (Labiatae) and *Smithorchis* (Orchidaceae). Some others of them are palaeoendemics which survived in this region, such as *Salweenia* (Leguminosae), *Sinadoxa* and *Tetradoxa* (Adoxaceae) and *Acanthochlamys*. Among them, *Acanthochlamys* – a relict genus – has been promoted to be a new family recently, (Wu and Wu 1998). Apparently, the tectonically active, deep cut and elevated geography has made the Hengduan Mountains not only an important centre of survival, but also one of the famous regions of speciation and evolution in the world, (Axelrod et al 1998).

Hong Kong

Hong Kong consists of a peninsula projecting from Guangdong together with an archipelago of some 230 islands occupying a total land area of about 1071 km^2. It is a paradox. It contains in its conurbations the highest density of human habitation anywhere in the world and yet 40% of its surface is designated a national park, consisting of thousands of hectares of grassland, scrub and woodland. Given the shallow surrounding sea, during the Ice Ages the Hong Kong islands would have been contiguous with the mainland. As a former British colony for many years, it was under the direct influence of Kew and the types of many species were described. Indeed, as early as 1861 Bentham produced *Flora Hongkongensis*.

Due to the efforts of many people there is now a comprehensive modern data base for the *Flora of Hong Kong* (Anon.) summarised in Table 3.4

Table 3. 4. Genera represented within families for Hong Kong (Spermatophyta only)

Genera /family	1–5	6–10	11–15	16–20	21+	Totals
Gymnosperms	5	-	-	-	-	5
Angiosperms						
Dicots	109	18	6	1	5	139
Monocots	35	2	-	-	3	40

Among the angiosperms the most conspicuous families numerically are given in Table 3.5

Table 3. 5. The more numerically conspicuous families in Hong Kong

Dicotyledons		Monocotyledons	
Acanthaceae	17/22	Cyperaceae	21/141
Compositae	64/105	Orchidaceae	54/125
Euphorbiaceae	26/65	Poaceae	97/221
Labiatae	24/37		
Leguminosae	55/132		
Rubiaceae	32/68		

It is instructive to note the similarities between this and Table 3.2 for Xishuang-banna for example.

Space does not permit detailed treatment here of all the families in Table 3.5, but it is worth pausing to consider the Orchidaceae. The following is based on the compilation of Siu Lai-Ping (2000). Of the 125 species, some 25% are epiphytic, a feature common among tropical orchids. Seventeen species are endemic to Hong Kong, including both epiphytic and terrestrial examples.

Commonly, a particular genus is committed to being either epiphytic or terrestrial but exceptions do occur. The genera *Bulbophylum*, *Eria* and *Liparis* contain examples of both sorts. *Aphyllorchis montana* and *Eulophia zollingeri* are terrestrial and saprophytic. Many of the orchid species are now rare or very rare – a fact which can be put in context by considering Hong Kong's vegetational history.

Forest Cover and Settlement

Given the temperature and rainfall regime, the pre–settlement vegetation would have been monsoon forest. Settlement began about 3000 years ago, leading to steady diminution of the forest cover. There are, however, some probably relict woodlands of considerable interest.

Feng Shui Woodlands

Deep in Chinese folklore is the notion of wind and water and the propitious placement of various objects. The validity of such notions are not our concern here, and a discussion of them we leave to others.

What *is* of interest is the existence of relatively undisturbed woodlands closely associated with small settlements. Within such woodlands is a non-random sample of the Hong Kong flora. Of the 179 angiosperm families in Hong Kong, only 35 are recorded as being represented, and in nearly half of these there is more than one species per family. With perhaps three exceptions (*Anredera cordifolia*, Basellaceae and two species of *Canarius*, Burseraceae) all the species in the woodlands are indigenes but are a mixture of rare (or very rare) and common species. More interest attaches to the rarities, since their inclusion in these woodlands could ensure their survival hitherto. They include *Popowia pisocarpa* (Anonaceae), *Wrightia taevis* (Apocynaceae), *Ilex macrocarpa* (Aquifoliaceae), *Aristolochia thwaitesii* (Aristotochiaceae), *Euonymus longifolius* (Celastraceae), *Elaeocarpus dubius*, *E. nitentifolius* (Elaeocarpaceae), *Mallotus oblongifolius* (Euphorbiaceae), *Castanopsis kawakanii* (Fagaceae) and *Nauclea officinalis* (Rubiaceae). The case is sharpened, sadly, by the instance of *Aphananthe aspera* (Ulmaceae), where the last surviving specimen was in a Feng Shui wood and was recently felled. Clearly, there is a strong case for such woodlands being given official protection on scientific grounds as has advocated by Corlett (pers. comm.).

Interesting as these woodlands are, they represent only one aspect of the absorbing biodiversity of Hong Kong which, as may be readily recognised, is vulnerable to the expanding commercial development of the area. An imaginative response has been that of the Kadoorie Foundation in setting aside land for the regeneration of, as nearly as possible, replicas of the original forest cover together with its animal indigenes

Xinglong Nature Reserve, Gansu

Gansu, containing a part of the Silk Road leading eventually to Asia minor and the Levant, is dominated, as is much of northern China, by the Loess Plateau. Much of the province is semi-desert but on the higher regions beside the Silk Road in the Hoxhi corridor, for example, alpine meadow occurs with *Gentiana* and *Potentilla*.

Xinglong Mountains 56 km south east of Lanzhou are part of the Qing and Qilian mountain area. The two peaks of Xinglong rise to more than 2400 m. Most rainfall is from April to September peaking in July-August. The mountains are tree–covered and with significant understoreys of vegetation. Table 3.6 summarises the seed plant families by genera.

Table 3. 6. Genera represented within families for Xinglong Mountain, Gansu. Data from Wang, undated)

Genera/Family	1–5	6–10	11–15	16–20	
Gymnosperms	3	-	-	-	3
Angiosperms					
Dicots	57	3	5	1	66
Monocots	9	1	-	-	10
	69	4	5	1	79

By comparison with Table 3.1, it is evident that the vegetation here is far less diverse but it is also sharply different in character being strongly temperate. (There is a prolonged period of winter cold lasting about 50 days). The tree cover includes *Betula, Corylus* and *Ostriopsis* (Betulaceae)*;* *Euonymus* (Celastraceae); *Quercus* (Fagaceae)*; Crataegus, Prunus, Pyrus* (Rosaceae) and *Tilia* (Tiliaceae). What distinguishes this from more European situations is, for example, the occurrence of three species of *Juniperus*, four of *Berberis*, seven of *Lonicera*, four of *Euonymus*, six of *Artemisia*, five of *Gentiana* and so on, testifying to an inherent richness in the vegetation. What is of particular interest is the conspicuous difference in this vegetation from that of the Loess Plateau nearby. Clearly, as a protected reserve, the dense tree cover is to some extent explained. What, however, deserves serious consideration is whether or not much of the vegetation is relictual and representative of what grew in areas previously which the Loess Plateau now covers. As yet the fossil record is not particularly informative on this point.

These are but a small fraction, though a significant one, of the immense range of the plant associations to be found in China and for which we can do little more than indicate some contrasts. We do, however, from different points of view, offer a further perspective on Chinese plant diversity in subsequent chapters.

4 From Domestications to the Arrival of the Western Plant Collectors

The earliest site anywhere in the world for agriculture appears to be Jericho in the Near East, and dating from about 10000 years ago. It was the subject of a classic study by Kenyon (1978). If that were indeed the earliest site, a problem which surfaces automatically is the relation of other sites to it. Are they derivative or themselves of independent origin? The further the distance from Jericho, the greater the likelihood of independence – completely so with the earliest sites in the New World. For China one might be less certain. Indeed, if it were shown that the earliest sites there preceded Jericho, the problem of dependence or not is not removed but merely reversed. On balance, the most realistic starting point might be that the original migrations of *Homo sapiens* out of Africa preceded by many millennia the cultural changes which led to agriculture. Relatively small, widely dispersed hunter-gatherer populations would have a life style and life expectancy dominated by the short term. Settlement, population increase, some degree of social control and sophistication would have to be in place for one human group to interact significantly with another. Until, therefore, compelling evidence emerged to the contrary, one might propose China as a third independent centre of origin for agriculture. Bray (1984) regards 7000 years B.P. as realistic for the start of agriculture in China with perhaps separate traditions for the north and south.

Nature of the Record

As a generalisation, writing, dating from Sumeria in 5000 B.P. is about half as old as agriculture. Chinese writing developed differently and is variously thought to date from 1800 to 1200 BC. In a form that can be read today, Chinese script became fixed between 221 and 206 B.C. As one moves from B.C. to A.D. at both ends of the Old World, literature on agriculture increases, but for the first millennium thereafter China produced more and better treatises than Europe (Bray 1984). However, for the origins of Chinese agriculture, as for elsewhere, the matter is archaeological.

Table 4.1. Historical development of plant agriculture in China

Approximate date	Location or event	Reference
6,000 BC	Hupei, rice	Smith (1994)
5,500	Shaanxi, millet	"
4,000	Gansu, millet, hemp (Yang Chao culture)	Bray (1984)
700-400	*Hsia Hsiao Cheng* (an early agricultural calendar)	"
400	*Kuan Tzu* *Book of Master Kuan* (reference to soya bean)	"
306	*Chhu Tzu* *Songs of the South* (First literary reference to sugar cane)	"
126	Supposed introduction of broad bean and pea from Central Asia	"
30 A.D.	*Chhien Han Shu* (Earliest bibliography of Agriculture)	"
535	*Chhi Min Yao Shu* *Essential Techniques for the Peasantry* Contains references to brassicas as oil seeds and details of forest management	"
550	*Lei Ssu Ching* *The Classic of The Plough*	"
770	*Chha Ching* *The Manual of Tea*	"
1273	*Nung Sang Chi Yao* *Fundamentals of Agriculture and Sericulture*	"
1313	*(Wang Chen) Nung Shu* *Agriculture Treatise of Wang Chen.* An important and comprehensive document	
1511	Earliest record of maize in China	"
1538	**Chung Yü Fa* *Treatise on Tuber Cultivation* (First mention of the peanut in China)	"
1639	*Nung Cheng Chhuan Shu* *Treatise on Agriculture*	"

*The author Hsü Kuang Chhi (1562 - 1633) was an early Jesuit convert and indicates the kind of east-west intellectual exchange which was opening up by then.

The Beginnings of Chinese Agriculture

Given the Chinese propensity to eat fish, whether from the sea or inland waters, the plants growing in inundated conditions could hardly have escaped notice. If seasonal flooding and drying out were then associated with the life cycle of annual grasses – those more likely to generate large quantities of seed, the scene would be set for a cultural shift in the direction of "managing" such a situation. This assumes that pressure of need from an enlarging population was in place.

Literary Records

Table 4.1 sets out a framework for understanding the development of agriculture in China. As is evident, literature begins to be generated early, though still thousands of years after the dawn of agriculture. The list of documents here is not complete but provides instances of significant milestones in the story.

Rice *Oryza sativa*

Cultivated rice, deriving from bamboo–type wild grasses, is a cereal important to much of the world's population. It shows its highest level of genetic diversity in the area from Assam to Yunnan. While therefore this might be to be presumed the area of domestication, the evidence for China of the earliest sites so far discovered is much further east. Do we assume there remain therefore earlier, more westward sites, to be discovered, or that some other kind of influence was at work?

The varietal diversity of rice is considerable – hardly surprising in view of the range of environments in which it is grown. It is, however, possible to allocate this diversity to subspecies of cultivated rice namely *indica, japonica* and *javanica*. *Indica* is adapted to a monsoon climate and accounts for most of the crop. *Japonica* is more temperate and *javanica* more equatorial in distribution.

Relationships among the three sub species have been the subject of considerable study and speculation since *indica* and *japonica* types, particularly, cross only with difficulty and the existence of such a sub–specific barrier is surprising. Given that the subspecies tend to be concentrated in contrasted environments, the simplest explanation has been that each represents a separate domestication from a widespread common ancestor – itself showing regional variation. Oka (1988) assuming that the wild grass *O. rufipogon* was the common ancestor, denied that *indica* and *japonica* types were represented there, and considered their divergence to be due to domestication. If this is so, it too is surprising in view of the relatively short time span involved.

A distinction is made between upland (dry) and paddy (wetland) rice, the bulk being of the latter kind. In fact, upland rice, too, shows improved yield if there is some waterlogging during its life cycle. Rather than a sharp distinction, it is more realistic to recognise a gradation in growing conditions between upland and paddy rice depending on resources available to the farmer, paddy conditions being preferred. Table 4.2 indicates significant events in the development of rice agriculture.

Table 4. 2. Emergence and development of rice agriculture

Approximate date	Location	Reference
8500 B.C.	Hupei, China	Smith (1994)
4500	Ho-Mua-Tu, China	"
3500	Thailand	Bayard (1970)
2500	Pakistan	Andrus and Mohammed (1958)
2300	India	Allchin (1969)
	" (carbonised grains)	Vishnu – Mittre (1974)
2000	" (cultivated)	"
2000	N. Vietnam (cultivated)	Chang 1976
1500	Indonesia, Malaysia and the Philippines	Spencer 1963
100 A.D.	First distinction made between *Hsien* (*indica*) long grained and *keng* (*japonica*) round grained rice	Shuo Wen Chieh Tzu quoted by Bray (1984)
535	Recognition of varieties including 12 non-glutinous and 11 glutinous varieties	Chhi Min Yao Shu Bray, (1984).
1012	30,000 bushels of Champa rice distributed to Lower Yangtze and Huei provinces from Fukien	Bray (1984)
1175	First reference to fragrant rice	"
1742	Recognition of 3000+ names representing about 1000 varieties	*Shou Shih Thung Khao* (quoted by Bray)
1904	First systematic hybridisation of rice by Kano in Japan	Grist (1951)
1962	Foundation of the International Rice Research Institute, Los Banos, Philippines	*Enc. Brit.* (1999)
1965-70	Development of IR8 from hybridisation of two *indica* rices Peta and Dee Gee Woo Gen (a significant contribution to the Green Revolution.)	Hardt and Capule (1983)

If, then, a search were made for the earliest sites with agricultural artefacts or plant remains and these did, in fact, occur in inundation–prone habitats then this would point to a rice–type original agriculture. So far such a process has led to the recognition of the Ho–mu–tu site across Hang–chon Bay some 160 km south of Shanghai dating from about 7000 B.P. – among the earliest sites so far discovered. More recently, evidence of rice culture has been dated even further back to about 8500 B.P. (Smith 1994). According to Kong et al (1996), evidence of rice culture 8000 B.P. has been found at two sites in Henan province, Jia Lake and Bancun.

In northwest China there appears to be a separate agricultural tradition, – the Yang-shao culture based on *Setaria italiaca* and *Panicum miliaceum* dated to about 7500 B.P. The settlements excavated are near the confluence of the Wei Ho and Yellow Rivers as they descend to the east China plain (see Bray 1984).

The Future of Diversity

Agriculture, worldwide, confronts a paradox. This is that given more widely dispersed forms of modern technology and the passing of germ plasm control into the hands of fewer, larger companies, the diversity of agriculture decreases. At the same time, genetic improvement for the crops which are retained technology depends on the widest possible trawl of hereditary factors controlling productivity. An awareness of this was put strikingly by Wilkes (1990):

"Suddenly, in the 1970s, we are discovering Mexican farmers planting hybrid corn seed from a midwestern seed firm. Tibetan farmers are planting barley from a Scandinavian plant breeding station and Turkish farmers planting wheat from the Mexican wheat programme. Each of these classic areas of crop-specific genetic diversity is rapidly becoming an area of seed uniformity"

Among cereals, modern technology is concentrated upon maize, rice and wheat, and to a lesser extent upon barley and sorghum and hardly at all upon a host of so called minor cereals. Yet to sustain high–quality breeding effort upon oats, pearl millet and t'ef is to hold, for human kind, a sort of insurance policy in case of failures, for whatever reason, in any of the three major cereal species. Publicly funded science has recognised this in the wide context global botanical awareness with, for example, the Kew seed bank based at Wakehurst Place, Surrey, in the United Kingdom and other similar institutions around the world. Genetic conservation has now developed its own orthodoxy to the point where it has become the customary, even routine material of so many plant breeding courses. It is therefore stimulating to explore Chinese classical literature and find an alternative approach to the problems of food shortage and famine.

The *Chiu Huang Pên TShao*

Given the climatic range of China and year–to–year variation in weather conditions, together with the kind of episodes indicated in Table 2.3, the occurrence of periodic food shortages and famines were features of Chinese history. Such a situation was not, of course, peculiar to China and other agricultural economies around the world had recourse to "famine foods" – the less convenient or palatable alternatives consumed when traditional crops failed. Matters in China took a different turn in the 15th century in a way both interesting and instructive.

The significant imperial figure was the Ming Prince Chu Hsiao (Ca. 1360 – 1425). He created a botanical garden and experimented with more than 400 kinds of plants. Eventually, details of these alternatives to standard crop plants, together with illustrations, were compiled in a book whose title opens this section and whose translated title is *Treatise on Wild Food Plants for Use in Emergencies*, published in 1406. Since the book concerned plants not previously used for food, if it were to serve its intended purpose it had to include recommendations for detoxification. One example, perhaps rather extreme, quoted by Lu and Huang (1986) will suffice. It concerns *Phytolacca acinosa*, containing what we now recognise as phytolaccatoxin and able to cause paralysis. It occurs in red and white forms and only the white should be used. The roots are sliced and washed repeatedly, preferably, it is recommended, by putting the slices in a basket and leaving in a stream for 48 h. It goes on to recommend that the slices be steamed and eaten with garlic.

This approach to famine foods has been called the Esculentist movement. Read (1946) translated the list of famine foods and attempted to attribute to as many as possible of them modern Linnean binomials. Appendix 1 contains Read's list together with various additional comments.

There is, in Western science, an echo of this approach with regard to oilseed rape, *Brassica napus*. Long known to have oily seeds, it was unsuitable as a source of cooking oil until plant breeders developed varieties almost free of erucic acid. There is, however, a further and perhaps rather more thought–provoking possible link with Western science.

N. I. Vavilov

One of the significant figures in 20th century plant science was Nicholai Ivanovich Vavilov, who drove the study of cultivated plants and their wild relatives through his numerous collecting forays in different parts of the world, together with his writings. He identified "centres of diversity" and, rather more controversially, sought to propound the view that they were "centres of origins of cultivation". As with any interesting innovator, he was provocative raising questions his successors have sought to answer. Of particular interest here is his interaction with Chinese agriculture. His list of Chinese plants includes 14 cereals and other grains, 3 bamboos, 19 thickened taproots, root tubers, bulbs and aquatic plants, 24 vegetables, 39 cultivated fruits, 1 sugar plant, 6 oil resin and tannin plants, 6 spice plants, 12 technical and medicinal plants, 6 fibre plants, 4 dye plants and 2 "miscellaneous" – in all 136. Neglecting the question as to what extent, given present–day knowledge, this is an under-estimate, Vavilov's conclusion is worth noting. He writes

"In wealth of its endemic species and in the extent of the genus and species potential of its cultivated plants, China is conspicuous among other centres of origin of plant forms. Moreover, the species are usually represented by enormous numbers of botanical varieties and hereditary forms…"

(Translated by K Starr Chester 1949/50)

How far is it possible to connect Vavilov's findings with the Esculentist tradition prompted 5 centuries earlier by Prince Chu Hsiao? Reference to Appendix 1 shows that among the species identified by Vavilov no fewer than 33 (indicated by the prefix V) occur in Read's list. We must not suppose that everything here was consumed only after the publication of the *Chu Huang Pên Tshao*, but the overlap is surely interesting.

Ornamentals and Their Collectors

Prior to Linnaeus, plants had been brought from China to the West but clearly he provided a considerable additional stimulus. Beyond this, matters began to commercialise as owners of nurseries and various wealthy sponsors commissioned one plant collector after another. Around this has gathered a considerable literature, part autobiographical, part biographical. Its characteristics are entry into unknown, often spectacular landscapes, the privations involved, and in the process the discovery and collection of plants which have become the prized possessions of generations of gardeners. It is a useful and necessary literature since it underlines the fact that we have these plants as the result of often considerable risk. If one were tempted to think of the "romance" of plant hunting, its practitioners in the 18th, 19th and early 20th centuries needed to be not only physically robust but to have a singular, relentless dedication to the task in hand. That said, it remains true that it is an oft–told tale and need not be retold here. What is rather more useful is to tabulate a sequence of the principal collectors and an indication of the materials with which they were concerned. Table 4.3 is derived from Coats (1969) to which the student is referred. Her book combines readability with high adventure and has an extensive bibliography of both original and derivative literature. More recent texts are Lancaster (1989) and Lyte (1989), auto–biographical and biographical respectively. A review of the latter book by Stearn (1990) is especially interesting. For information by this last author on introductions from Japan, see Stearn (2000). Several plants of Chinese origin are illustrated in Plates 4, 5 and 6.

Table 4. 3. Important Western plant collectors In China

(After Coats 1969)

Cunningham, James Ca. 1698	First description of tea cultivation, 600 species recognised including the first ornamental *Camellia*.
Ternstroem, C. Ca. 1735	An associate of Linnaeus, commemorated by him in *Ternstroemia* (Guttiferae)
d'Incarville, Pierre Nicholas le Cheron (1706–1757)	*Incarvillea*
Osbeck, Pehr	An associate of Linnaeus, commemorated by him in *Osbeckia* (Melastomataceae)
Torin, Benjamin, Ca. 1770	*Cordyline terminalis, Daphne odora, Murraya exotica, Osmanthus fragrans, Saxifraga tomentosa*
Haxton, John, Ca.1792	*Camellia sasanqua* (*oleifera*)
Main, James (1770 - 1846)	Tree peonies, *Chimonanthus praecox, Clerodendron fragrans*
Staunton Sir George, Ca. 1793	7-8 novelties in *Rosa bracteata*
Kerr W, d. 1814	Introduced 700 living plants, including 238 new ones
Abel, Dr Clarke (1780–1826)	*Abelia chinensis*
Potts, John Ca.1821	*Camellia multiflora, Callicarpa rubella, C. japonica, augustata, Primula praenitens* = *sinensis*, 40 varieties (all lost) of *Chrysanthemum*.
Parks, John Damper Ca. 1823	20 - 30 *Chrysanthemum* varieties, *Rosa banksiae* double yellow, *Camellia reticulata* and the first *Aspidistra* (*lurida*).
Fortune, Robert (1812–1880)	*Caryopteris mastacanthus, Diervillea rosea* (*Weigelia*), *Forsythia* spp. *Mahonia bealli, Phalaenopsis amabilis, Skimmia reevesia* and many others.
David, Armand (1826–1900)	*Prunus davidiana, Rosa xanthina, Xanthoceras sorbifolium* and others
Farges, Paul Guillaume (1844–1912)	*Davidia involucrata* and *Incarvillea grandiflora*. He is commemorated in the bamboo *Fargesia*.
Delavy, Jean Marie (1838–1895)	He is credited with 1500 new species. His finds include *Incarvillea delavayi, Paeonia lutea* and *Meconopsis betonicifolia*.
Soulie, Jean André (1858–1905)	*Buddleia davidii*
Henry, Augustine (1857–1930)	Many finds including *Acer henryi* and *Rhododendron augustinii*
Wilson, Ernest Henry (1876–1930)	(Dug up 7000 bulbs) His finds include *Aconitum wilsonii, Actinidia chinensis, Magnolia wilsonii* and many others
Forrest, George (1873–1932)	*Gentiana sino–ornata, Pieris formosa*, several *Primula* species and numerous others
Meyer, Frank (1875–1918)	*Ginkgo biloba*
Kingdon-Ward, (1885–1958)	
Farrer, Reginald (1880–1920)	Mostly together with Purdom, William (1880 - 1921) Many acquisitions among them *Clematis* spp., *Gentiana farreri, Trollius pumilus* and *Viburnum farreri*

Part II

Growth Forms:
Some Representative Taxa

5 Trees

"All the available evidence points to the fact that the Flowering Plants, from the time they are first recognised... right down to the middle or end of the Pliocene, pursued the even tenor of their way... their history seems to have been that of... broadening and differentiating by the multiplication of forms. Then quite rapidly... during the Pliocene this idyllic sequence was broken by a drastic deterioration in the climates of higher latitudes, culminating in widespread glaciation and presenting to the world of Flowering Plants problems of environmental harmony which it had never before encountered".

Good (1964)

The effects of glaciation described above by Good were much less severe for China. If the idyllic sequence of which he writes, though perhaps modified, persisted there, it can help explain how for some species that country served as a place of last refuge. *Ginkgo biloba,* a gymnosperm, for example, is the sole survivor there of a genus formerly widespread in the world and with an immensely long fossil history dating back nearly 200 million years. Seward (1938) charts the decline of the gingkophytes – the close relatives of *Ginkgo.* By the beginning of the Oligocene only 2 out of 19 genera remained. The decline continued into the Miocene and they disappeared in the Americas. In Europe *Ginkgo* appears to have survived into the Pliocene and then succumbed to the effects of glaciation. Wilson (1929) attributed its survival in China to protection in Buddhist monastery gardens, but this appears to be unfounded. What are surely more relevant are the benign effects of having escaped the ravages of glaciation.

It is possible to extend this approach from *Ginkgo* to a range of trees, some ornamental, some of use in forestry, and in so doing understand their present distribution and additionally why they have aesthetic appeal or utility. Because in addition to gymnosperms, the angiosperms are important, here the genus *Magnolia* and some of its relatives are used to set the scene. In Chapter 6 and those following more attention is given to the range of botanical diversity. (*Ginkgo,* additionally, is considered in a medicinal context in Chapter 9).

A Brief History of the Magnolieae

A prerequisite here is to recognise that, in addition to present–day species, there are those of the same genus which have become extinct. In the Tertiary flora of England, for example, Reid and Chandler (1933) recorded *Magnolia angusta, crassa, lobata, subcircularis, subquadrangularis, subtriangularis* and *Talauma wilkinsonii*, though none would be found in a modern list of living species from anywhere in the world. This last resembles *T. angatensis*, a presently living species from the Philippines, suggesting that Herne Bay in southeast England, where the fossil form was found, once had a lowland tropical environment.

The Magnolieae is a convenient term to cover part of the Magnoliaceae and includes the genera *Magnolia, Michelia* and *Talauma*. (A fourth genus, *Manglietia* can be seen either thus or subsumed under *Magnolia*). Many years ago, Good (1925) described the 128 species then recognised as distributed through North and Central America, the Malay Archipelago, Himalayas, Burma and Indo-China and China with outliers in the West Indies, Japan and the Malay Peninsula.

If the Magnolieae are separated into genera, *Talauma* is mostly tropical, *Michelia* is primarily oriental, with *Magnolia* found in the more temperate regions outside of Europe and the Near East; but this relates only to the present day. In Tertiary times a more equable climate existed with a largely complete continuity of distribution across the Northern Hemisphere for the Magnolieae.

Good (1925) supposed that the advance of Pleistocene ice was relatively slow, allowing the Magnolieae over the generations to migrate southwards. Clearly, if, as in the Americas mountains run north–south, this is easier than if they run east–west as in Europe. Elimination in Europe according to this view, is far more likely than in North America, since in the latter case, the return of warmer conditions offers the possibility of reverse migration. This, in outline, would help explain the elimination of *Magnolia*, for example, from Europe and its survival in the south eastern United States and the Far East (though in these two cases for somewhat different reasons). In Idaho, in the northwestern United States, *Magnolia* does not nowadays occur naturally. However, Filgar (1993) has described from there well–preserved leaves and a fruit aggregate of *Magnolia* from about 17 Ma closely similar to present–day *M. grandiflora*.

Earlier, the Himalayan upthrust was described as a consequence of collision between India and Eurasia, and the results for the monsoon and the development of the Loess Plateau were indicated. The aftermath of continental drift and, during the Tertiary, the development of the present world mountain systems led to a zonation of climate, the differences between the equator and the poles becoming much more marked. Under such circumstances the onset of the Pleistocene glaciations distinguished between more and less vulnerable areas of vegetation. Oceanic islands at equatorial latitudes would have been the least affected. While the sea level would have decreased appreciably, nonetheless, such islands would have been buffered against extremes of heat and cold, then as now.

It is necessary following this simplified account to enter some reservations. The climate of the Tertiary was not uniform in time nor space. Local variations occurred. The continental landmasses were not stationary; and of course, not only would the three genera of the Magnolieae have different ecological requirements; this would also be true of the individual species of which they were comprised. Finally, present–day species presumably could have evolved from their Tertiary counterparts, and conclusions we make about their present–day ecological requirements need not match exactly those of their forebears, although it seems reasonable to expect at least similarity. What we do have from the Magnolieae and from *Magnolia* in particular is a useful guide in understanding aspects of plant distribution. Unlike the gymnosperms, however, *Magnolia* and other angiosperms cannot be traced back in time beyond the Cretaceous, the period preceding the Tertiary.

On this basis we now consider selected examples of Chinese tree genera, some interesting as ornamentals and some significant in the long history of forestry there. Not all, for our purposes, are of equal importance, and more attention is given to the most interesting cases.

Gymnosperms

Pinaceae

Abies

On a world basis there are about 50 species, of which 22 occur in China. Often with blue-grey foliage, they are handsome trees, some well_known as ornamentals with coloured cones. Some species are immensely tall and yield valuable timber. Because of problems with conservation of *Abies* in China, it is reconsidered in the concluding chapter. See Plate 7

Cathaya

Cathaya is a monotypic, rare and relict genus in the Pinaceae endemic to China. *Cathaya argyrophylla* was discovered in 1958 at Huaping, Longsheng County, Guangxi Province, China, by a group of botanists who were conducting fieldwork in the area, (Wang 1990). At first, a seedling looking like *Keteleeria* was found and then specimens with cones were collected. The plants were immediately recognised as being unique and a new, monotypic genus, *Cathaya* in the Pinaceae, published by Chun Woon-Young and Kuang Ko-Ren in 1962. See Plate 8

The discovery of cone fossils from Pliocene sediments east of the Black Sea, and from layers of the Miocene to Pliocene in the Aldan River Valley in E. Siberia, (Florin 1963; Ferguson 1967) indicates that *Cathaya* was widely distributed in Eurasia in the Tertiary. Based on these findings, *Cathaya argyrophylla*, Chun et Kuang, the only extant species, has been considered to be a living fossil. It was another exciting discovery in the history of botany after that of *Metasequoia*.

The strong climatic changes during the Quaternary resulted in a decrease in the number of individuals and the discontinuous distribution of *C. argyrophylla* populations. These events led to the low genetic diversity and the great reproductive barrier that brought about the decline of the species and the shrinkage of the distribution range to where *C. argyrophylla* is now endangered, (Florin 1963; Ferguson 1967; Wang 1990; Fu and Jin 1992).

Since the publication of the new genus, it has attracted the attention of many researchers in various fields, including taxonomy, palaeophytology, palynology, embryology and anatomy (Wang 1990). The leaf arrangements on long branches and the crown of *Cathaya* are very similar to those of *Pinus*, but leaves of *Cathaya* do not form bundles. There has been much discussion on the systematic position of *Cathaya* in the Pinaceae and its status as a genus, (Greguss 1972; Wang and Chen 1974; Hu and Wang 1984; Wang 1990; Chen et al 1995). The researches on embryology and molecular systematics further reveal that *Cathaya* is an independent and primitive genus in Pinaceae and that it is closer to *Pinus* than to *Pseudotsuga*, (Wang and Chen 1974; Wang et al 1997, 1998). Recent molecular studies reveal that *Cathaya* may have become established in the early and middle Cretaceous together with *Pinus*, *Picea* and *Pseudolarix*. The lack of early fossil records of the monotypic genus *Cathaya* may be due to their limited historical distributions and / or less extensive studies of fossils at these sites, (Wang et al 2000).

Cathaya argyrophylla as a rare, relict species endemic to China, is discontinuously distributed in the subtropical mountain areas. Its occurrence is very scattered, and colonies consist of from one to a few individuals, or at most several dozen, except at Laotizi, Jinfo Mountain, where the density is higher. The total number of individuals with a height greater than 1m is less that 4000. The soils where this species can survive are developed from limestone, shale and sandstone and are slightly acid. It grows along the crests of isolated cap-like rocky mountains and in crevices of sheer precipices and overhanging rocks with shallow soil. All existing colonies are not easily accessible. Flowering and pollination take place in May, fertilization occurs in June of the following year, and the cones ripen in October, (Xie 1996; Wang 1990).

As a rare and endangered species, *Cathaya* is listed in the China Plant Red Data Book. Seven *Cathaya* communities have been found in China, in which *Cathaya* is one of the dominant species. Natural reproduction of *Cathaya* depends, to a large extent, on the density of the forest. It regenerates well within thin stands, but its seedlings and saplings are rare in dense stands. Although shade is necessary for the seedlings, the older the tree, the more light is needed. *Cathaya* is often found mixed with broad-leaved trees and is shaded by them. The natural regeneration of *Cathaya* is thus very poor in these communities. It is threatened by the invasion of broad-leaved species and is likely to be replaced by them during the succession. This is currently an important factor limiting the enlargement of the populations of *Cathaya*. Low seed production and the low germination rate are other factors resulting in its endangered status, (Wang 1990;

Xie and Chen 1994; Xie 1996). Its genetic variability and genetic structures of populations have been detected by using allozyme profiles and RAPD, (Wang et al 1997; Ge et al 1997). Preliminary results show that *Cathaya* possesses the lowest level of genetic variability among gymnosperms and its genetic structure, with strong genetic differentiation not only between populations but also between subpopulations, is unique among conifers, (Ge et al 1998).

At present, every known population has been given full attention and placed under efficient protection by the local governments and departments concerned. Nature reserves (Jinfo Mountain and Wunong in Sichuan, Huaping and Jinxiu in Guangzi and Chengbu and Linxian in Hunan) have been established, and the emphasis has been placed on the protection of *C. argyrophylla*. In the arboretum in Xinning County, Hunan Province, *C. argyrophylla* is growing successfully and additionally, some grafted trees between *Cathaya* and *Pinus elliotii* are being cultivated successfully there too.

Pinus

The pines are a remarkably resilient group of gymnosperms which have not merely survived from before the rise of angiosperms but have radiated through a range of species to occupy tropical, temperate and cold forests. Chinese indigenous species include *P. sylvestris, bungeana, tabuliformis* and *massoniana*. See Plate 7. Logically, it makes sense to grow pines (and other trees) on the steeper slopes to protect watersheds, and this is commonly done in China. The pines for the most part produce a 'prosaic' timber – plentiful, fast-growing, easily worked and mostly unremarkable in both grain and usage.

According to preference, pines are celebrated for their beauty, or not, but clearly they are immensely useful. Whether or not they are appreciated aesthetically, they form a conspicuous item in the Chinese landscape. Soot derived from the burned branches is a traditional source of scholars' ink.

For a more detailed survey see Daniels (1996). *Pinus* as an item for conservation is referred to in the concluding chapter of the present book. For a convenient and authoritative treatment of Chinese conifer taxonomy see Wu and Raven (1999).

Taxodiaceae

Cunninghamia (Plate 16). This is primarily of importance for timber in southern China, where its sprouting from cut stumps is well known. Even as early as the *Erh Ya*, its qualities were appreciated. The tree is fast–growing and, if left, will make magnificent specimens perhaps 25 m or so tall, reminiscent in habit of American redwoods. However, a more likely outcome is that trees will be cut for boards after about 20 years. One species, *C. lanceolata* is widely known and another more recently discovered is *C. unicaniculata* (Wang and Liu 1982).

Metasequoia

One particular interest regarding this plant is that it was first known from fossil material and believed to be extinct for many millions of years. Subsequently, it was found living and could thus be regarded as a living fossil. Its recovery for science is therefore worth recalling.

In 1941, Miki described from Japanese material this new fossil genus and attributed to it two species, *M. disticha* (formerly *Sequoia disticha*) and *M. japonica* formerly *S. japonica*). No living relatives were believed to exist. Eventually this prompted a reappraisal of fossil records elsewhere including China. References are Hu (1946) and Li and Zheng (1995).

Based on Bartholomew et al (1983) regarding the discovery of *Metasequoia*, the following sequence of events is thought to have occurred. In 1941 at Modaoqui in Sichuan province near the border with Hubei province, living specimens of *Metasequoia* were found, though not initially recognised as such, but rather thought to be *Glyptostrobus* – another monospecific genus confined to China. Eventually, Cheng Wanjun of the National Central University in Nanking realised the material was *Metasequoia*, hitherto known only from the Pliocene fossils described by Miki earlier. In 1947 seeds were sent to botanic gardens in America (Arnold Arboretum) and to Copenhagen and Amsterdam. Seed also arrived in Cambridge at about the same time. What was a botanical sensation half a century ago is now a fairly commonplace, though very attractive, item available from garden centres. Additional details are available from Bartholomew's account.

Again, as with *Ginkgo*, China remained, for *Metasequoia*, the place of last refuge. For reasons now obvious, the name adopted by C J Hsueh (Jia-Rong Xue) for the type specimen of the living material is *M. glyptostroboides*, more widely known now as dawn redwood. Quite apart from its obvious horticultural interest, absorbing botanical questions are prompted by the long fossil, and now living, history of this species.

Chaney (1948) reexamined American fossils formerly assigned to *Sequoia* which he believed should, in many cases, be reallocated to *Metasequoia* (these fossils reaching far north to latitude 82°). He also observed that this genus is *deciduous*, consistent with the majority of angiosperm associates in the Arcto-Tertiary flora. It then becomes possible to compare the fossil flora associated with *Metasequoia* with the living material accompanying it in present–day China. More recently, Bartholomew et al (1983) took a related but recognisably different approach. They noted that the natural *Metasequoia* habitat in China recalled that of *Taxodium* (swamp cypress) in America. Pooling the results from several studies, we find the situation as set out in Table 5.1. A glance down the table reveals evident similarities but with what might be called characteristic American and Chinese flavourings. Beyond this the following points are noteworthy.

1. Fifteen genera are shared. *Carya*, in this instance, is arguably replaced by *Pterocarya*, although the former genus occurs elsewhere in China.

2. (Not included in Table 5.1), *Saururus cernuus* in the United States is, arguably, "replaced" by *Houttuynia cordata* in China in this situation. Both belong to Saururaceae.

Given that evident similarity exists between the "*Taxodium* flora" and its *Metasequoia* equivalent, what significance attaches to this? Were only one or two associates involved, the matter might be dismissed as a matter of chance. As the number of associated genera increases, we ask whether this could represent some durable phytochorion traceable back to now fossilised Tertiary floras or if it represents but one aggregate of the many that have assembled and dispersed in

Table 5. 1. A comparison of *Taxodium* and *Metasequoia* associate species (After Bartholomew et al).

S E United States	S W China
Taxodium distichum	*Metasequoia glyptostroboides*
+	+
Acer rubrum	*Acer* spp
Ampelopsis spp	
Berchemia scandens	*Berchemia* spp
Betula nigra	*Betula luminifera*
Bignonia capreolata	
Carpinus caroliniana	*Carpinus fargesii*
Carya spp	
	Clethra fargesii
	Cocculus orbiculatus
Cornus spp	*Cornus controversa, macrophylla*
Decumaria barbara	
Fraxinus spp	
Ilex spp	*Ilex* spp
Itea virginiia	
Liquidambar styraciflua	*Liquidambar acalycina (formosana)*
Lindera benzoin	*Lindera glauca*
	Morus sp.
Nyssa aquatica, sylvatica	*Nyssa sinensis*
Parthenocissus quinquefolia	
Populus heterophylla	*Populus adenopoda*
	Pterocarya hypohensis, paliurus, stenoptera (Plate 4)
Quercus spp	*Quercus* spp
Rhus radicans	
Salix spp	*Salix* spp
Smilax spp	*Smilax* spp
	Styrax bodinieri, japonica, suberifolius
Ulmus americana	*Ulmus multinervis*
Viburnum spp	*Viburnum* spp
Vitis spp	

response to migration and changing circumstances. For further discussion of these ideas see Chaney (1948) and Bartholomew et al (1983). Some other ingredients in Table 5.1 are interesting in their own right and are considered later.

One concluding point is of interest here. After *Metasequoia* was (re) introduced to the West in 1947, its progress in various botanic gardens was closely followed and among other things to emerge were the wide differences in frequency of cone formation. For American and British accounts see respectively Wyman (1968) and Mitchell (1970)

Taiwania

This genus, another confined to China, is represented in three places sufficiently separated as to constitute a disjunct distribution – namely Manchuria, Taiwan and Yunnan. One might suppose a land bridge formed during a world glacial phase allowing the appropriate migrations between the first two. Its occurrence on the opposite side of China might be explained by supposing elimination of a once more or less continuous forest. Fossil evidence is not available on this point. Transfer by human agency seems hardly likely and the matter remains perplexing.

Further light might be shed by a consideration of the specific status of Taiwan and Yunnan representatives. One view is that two species are represented, *T. cryptomerioides* to the east and *T. flousiana* to the west. If their distinctiveness is convincing, the argument against human transfer would be strengthened, as would be the case for long–term regional diversification.

In the meantime, it appears that this impressive conifer, able, in some individuals, to survive 2000 years, and reminiscent of *Sequoia* in general aspect, remains relatively unknown in the gardens of the West although specimens of *T. cryptomerioides* are growing well at the Bedgebury Pinetum in southeast England, for example. In China, *T. flousiana* is classified as a "first grade, state protected plant". Fajon (pers. comm.) takes the view that the genus has only one species and that "*T. flousiana*" as a separate species is unjustified.

Evolutionary Divergence: a Novel Approach in Conifers

Before leaving these conifers, attention is directed to a recent study confined to the Pinaceae but with obvious implications for the other coniferales (Araucariaceae, Cephalotaxaceae, Cupressaceae, Podocarpaceae, Taxaceae and Taxodiaceae). It depends on these considerations:

1. The chloroplast genome is inherited paternally
2. The mitochondrial genome is inherited maternally
3. The nuclear genome is biparental

4. For each organelle considered across genera can show degrees of divergence attributable change through time.
5. By considering each organelle separately and then collectively a 'tripartite' view of conifer evolution can be explored using the constituent DNA.

Wang et al (2000) using this approach with 11 genera in Pinaceae, showed, for example, *Pinus* and *Pseudolarix* widely diverged with *Cathaya* nearer *Pinus*. Thus, the DNA assessment tended largely to confirm taxonomic judgements made in more traditional ways.

Angiosperms

From the vast number of tree genera available, the aim here has been to select examples on the basis of the issues they raise rather than to provide a comprehensive and, perhaps, relatively dull catalogue. The following instances provide a means of illustrating endemism, evolution, the importance of location, species numbers, hybridisation in nature, the relevance of traditional Chinese botany, the postglacial "restocking" of the European flora and the role of forestry in modern China, all themes which recur elsewhere in this book. The examples chosen are not completely incidental, but many others might have been substituted.

The dicotyledons, for interest, are presented as examples of Archichlamydeae and Metachlamydeae – technical terms for free and fused floral parts, respectively, and implying more "primitive" and more "advanced" floral structures.

DICOTYLEDONS

ARCHICHLAMYDEAE

Bretschneideraceae

The family is monotypic, with one species, *Bretscheidera sinenis,* almost confined to southern China though extending into northern Vietnam. It is best regarded as a relict genus formerly more widely spread, and is dioecious. It occurs in broad-leaved evergreen and deciduous mixed forests. Bretschneider was an early and important European student of Chinese plant life (see, for example, Bretschneider 1898).

Calycanthaceae

Legend has it that the Greek goddess Athena sprang fully formed from the head of her father, Zeus. From the fossil record, one has, almost a similar impression of flowering plants. Their appearance is relatively sudden and without convincing close antecedents.

One response among botanists has been to search for "living primitives" or "living fossils". By these are meant, in this context, existing flowering plants in, which are concentrated, a mass of supposedly primitive features. No family combines all such features, but there are several families which each have, to some extent, different arrays of primitive features. Behind this is the notion that clearly conjectures the nature of the original angiosperm stock and its immediate forerunners. Together with the Magnoliaceae, these families include the relatively obscure ones such as Degeneriaceae, Eupomatiaceae, Lactaridaceae, Himantandraceae and some others, among them Calycanthaceae, having two genera, *Calycanthus* and *Chimonanthus*, the latter endemic to China although both occur here. In common with other similar families the floral parts are spirally arranged with no calyx/corolla distinction, the carpels are separate, and the anthers have short filaments with a broad connective. Pollination is by beetles. As indicated earlier, these families are not wholly primitive and, in Calycanthaceae, reduction of the carpel to an achene-like structure, the concave receptacle and seeds with relatively large embryos and no endosperm at maturity, would be considered advanced features.

Of *Chimonanthus*, there are three species, *C. praecox, nitens* and *salicifolius*, which occur more in central and eastern China. The first, *C. praecox*, is widely cultivated in temperate gardens as wintersweet, its perfume enlivening bleak January days.

Eucommiaceae

The family is endemic to China and has one species, *Eucommia ulmoides*. Since medicinal properties are attributed to the bark, *Eucommia* is cultivated near dwellings and occurs from central to eastern China. The specific epithet is interesting. The family is perhaps related to Ulmaceae.

Fagaceae

Quercus

How many species of oaks are there? On a world basis, estimates vary. Good (1964) speaks of "between 500 and 1000". More recently Daniels (1996) refers to "800 or more" adding ambiguously "including many hybrids". Within a Chinese context this latter author quotes Lee Shunch'ing in 1935 with 54 species and in 1973 with 164. We really do not know how many oak species there are in the world, let alone in China. Wu and Raven (1999) suggest 300 on a world basis, many of which are known in China, including 15 endemic species.

Traditionally, in China, in addition to timber, oaks have provided forage, acorns, as a source of both food and dyestuffs, bark cork, fuel and tannin. Perhaps surprisingly to a Westerner, oak leaves, too, can feed silk worms. Daniels' comment (1996) has recently found a new and surprising significance in work originally done in Europe. The essentials are as follows:

1. In northwest France *Quercus robur* was the pioneer species, following glacial retreat due to rare, wide-ranging founder events.
2. Following these came local diffusion
3. With increasing forest density *Q. sessilis* tends to be at competitive advantage
4. The overall impression is that two supposedly different species of oaks are in intimate genetic relation, one consequence of which is that *Q.sessilis,* though of later entry reaches ecological dominance "on the back of" *Q. robur.*

For a full account see Petit et al (1997).

Among the oaks of China we should, surely not rule out the possibility of similar inter relations among different species, although no current investigations are known to the present authors.

Salicaceae

Populus

In names which have tended to pass out of botanical usage, the "Amentiferae" (Amentiflorae, Amentaceae) subsumed several unrelated families, notably Betulaceae, Fagaceae, Juglandaceae, Leitneriaceae, Myricaceae and Salicaceae. Typically, they produce catkins and are wind–pollinated. Hybridisation is common among related species and creates taxonomic problems, relevant here, in *Populus, Quercus* and *Salix.* Statements therefore about how many species of these genera occur in China, or elsewhere, need to be treated with reserve. However Wu and Raven (1999), reckon on 100 species of *Populus* worldwide, of which some 47 are endemic to China.

Populus species are evident today throughout much of northern China. *P. tomentosa,* deliberately planted in the Hoxhi Corridor, shows evidence of salt damage through the use of poor–quality irrigation water.

Salix

Apart from bamboos, perhaps no tree genus is more thought to characterise China than the willow.[1] There is evidence that 400 years before Christ willows were recommended for stabilising dykes. The *Chhi Min Yao Shu* recognises four species of willow. Where bamboo is not readily available, osier–willow is used for basket making here as elsewhere. Otherwise, its timber is seldom that of first choice.

From a modern viewpoint, defining willow species and hybrids is a problem, in China as elsewhere, for reasons alluded to earlier. One estimate, Lee Shun-neh'ing (1973) quoted by Daniels (1996) is 173 species for China. More recently, Wu and Raven (1999) estimate 530 species worldwide with 189 Chinese endemics. Those species, both the endemics and otherwise in China, form a complex group presented in 36 sections of the genus. Although catkin-bearing, flowers of willow have nectaries and are entomophilous.

Ulmaceae

Ulmus

Some 40 or so species of elm exist worldwide, spread across the milder parts of the Northern Hemisphere. Probably five species, *U. pumila, davidiana, macrocarpa, uyematsuii* and *changii*, are significant in China. Traditionally, in the West, elm has been regarded as a low–grade timber liable to warp, and suitable for farm carts, cheap furniture where strength accompanied by a tendency to slight warping was of little consequence, and for coffins. Technology based on extended drying times has come to the aid of elm in England, and for about 60 years it has been used with remarkable success to make very high–grade furniture, where the attractive grain is shown off to great advantage.

One curious use of elms historically in China has been to plant it to make defensive bulwarks.

---

[1] The "willow pattern", seemingly Chinese with weeping willow, pagoda-like structures and three men on a quaint bridge with two lovers turned into swallows is, in fact, thoroughly English. It was developed by Thomas Turner and Thomas Minton between about 1779 and 1793.

METACHLAMYDEAE

Bignoniaceae

Catalpa

Directions for planting *C. ovata* from seed in the autumn and transplanting seedlings after five months are given in the *Ssu Min Yüeh Ling* (Table 5.2). This species yields a hardwood timber and is mostly confined to southern China. There are perhaps 11 species in total occurring naturally in east Asia, America and the West Indies of which four are known in China *C. bungei, fargesii, ovata and tibetica,*. Wu and Raven (1998). *C. bignonoides* is a well–known ornamental of Western gardens and will hybridise with *C. ovata*, being made available as *C* x *teasii* named after the American nursery of J C Teas which originated it.

Nyssaceae

Davidia

Named after Armand David, this tree is distinguished by two large bracts which subtend its inflorescence, hence its common names of dove or handkerchief tree. The genus has one species and that is confined in nature to China although the tree is planted widely through Western gardens as an ornamental. For a comment on Davidiaceae see Chapter 14.

Whether, in common with other genera in its family, it formerly had a wider distribution is uncertain, though perhaps likely. Alternatively, its weird flowering structure might conceivably have evolved within China. It had to happen somewhere.

Nyssa

Reference to Table 5.1 shows the existence of *Nyssa* species associated with *Taxodium* and *Metasequoia*. It is worth mentioning that the Tertiary flora of Europe contained several species of *Nyssa* known from fossils, although the genus does not nowadays exist naturally there, suggesting the kind of recession illustrated previously by *Magnolia*. *Nyssa* provides two familiar ornamentals, *N. sinensis* and *N. sylvatica,* grown for their spectacular autumn colouring.

The genus *Liquidambar* (Altingiaceae or Hamamelidaceae), though unrelated to *Nyssa*, closely parallels its distributional history. At present it does not occur naturally in Europe although the now extinct *L. palaeocenica,* for example, occurred in the Tertiary flora of southern England. *L. styraciflua* is also a source of autumn colour.

Oleaceae

Syringa

Although Oleaceae occur widely, several genera, notably *Syringa,* are largely confined to China, and a consideration of this genus offers an interesting opportunity to reappraise a garden ornamental, taken largely for granted, in terms of varieties of *S. vulgare* grown commonly.

There are perhaps 20 species of lilac reaching from south east Europe eastward to Japan. Among these 16 occur in China of which 12 are endemic. Of these endemics no fewer than ten are represented in Sichuan. There is also a significant representation of *Syringa* in the northeastern provinces toward Korea. In terms of accessibility to Western gardens many Chinese lilacs are available, for example, via the *Plant Finder* (Lord 1997–98) and some, such as *S. konarovii* and *S. pubescens,* have contributed outstanding variants.

Nowadays, the choice for the horticulturist is to balance the prospects of hybridisation against those of new collections from the wild, as the most likely source of new varieties. *Syringa* as an item for conservation is noted in Chapter 15.

Scrophulariaceae

Paulownia

Paulownia is a genus of seven species native to east Asia of which six occur in China. Although placed in a separate family to *Catalpa,* it is recognised that the two genera stand close to the common stock of the two families. *P. fortunei* and *P. tomentosa* provide timber traditionally regarded as giving an excellent quality of sound to musical instruments, (*Chhi Min Yao Shu* 535 A.D.). At the surprisingly early date (by Western standards) of plant monographs, in 1049 the *Thung Phu* confined to *Paulownia* was published. One species *P.* x *taiwaniana* is thought to be a naturally occurring hybrid *P. kawakomii* and *P. fortunei,* Wu and Raven (1998)

MONOCOTYLEDONS

Arecaceae

Caryota

For reasons explained earlier in the discussion of *Magnolia*, the band of palms encircling the earth equatorially, many of them on islands, is unlikely to have been affected radically through successive phases of glaciation.

This genus, consisting of some 12 species, reaches from Sri Lanka to northeast Australia and has a northward extension into southern China. Here, *C. ochlandra* makes a tall tree, its inflorescences being some 3 m long. Of the other species outside of China *C. mitis* and *C. urens* are used as ornamentals in the southern United States. *C. mitis*, the shorter species rising to 8 m is, feasibly, a glasshouse subject.

The leaf form, unique in this family, gives them the popular name fish–tail palms.

Trachycarpus

This genus provides an instructive contrast to the previous one in terms of cold adaptation. Three genera *Chamaerops, Rhapidophyllum* and *Trachycarpus* are unusual among palms in showing a degree of cold adaptation – taken to its extreme in the last of these. *Chamaerops* is found in Spain and Portugal, and *Rhapidophyllum* in the southeastern United States.

Trachycarpus on present reckoning occurs as eight species in Himalayan and east Asian regions although, in past times, *T. raphifolia*, now extinct, is known to have occurred in the Tertiary flora of southern England. Lancaster (1989) made the interesting observation that although *T. fortunei* occurs in the Yangtze gorges and widely elsewhere in China, Robert Fortune's introduction of it to the West (not, it may be noted, the first such) in 1849 was from Chusan Island (Zhoushan) off the coast of Zhejiang Province. Conceivably, it owes its present wide distribution throughout China to human agency, since the plant is credited there with medicinal properties and its fibres are useful. Under the circumstances, it would be difficult to support the notion that Chusan (or any other area of China) provided a nucleus population.

Since *T. fortunei* is remarkably cold–tolerant this, together with, say, *Hibiscus syriacus* and *Canna indica*, planted beside southern English swimming pools, might help create at least the illusion of a climate more benign than is actually the case.

Poaceae

Bamboos

These, of which there are numerous genera, provide a highly versatile material for a vast range of uses, as is well known. Bamboo strands, if sufficiently tightly woven, can even make a bucket to carry water. A modern treatment of bamboo is Chapman (1997), where the group is described from a range of aspects including the properties of its timber. The bamboos are considered briefly in the following chapter.

A View from the Trees

Perhaps surprisingly in this chapter, both ornamentals and more utilitarian trees have been considered together. A major theme has been the relative impoverishment of the European flora at the close of the Pliocene and into the Pleistocene.

With the last glacial retreat, revegetation began, and among the tree species elm, oak, poplar and willow are predominant, and it is no accident that all of them appear in Table 5.1. Europe, left to nature, did not reclaim all its treasures, but these four durable genera ensured that across the Northern Hemisphere, in temperate locations, the deciduous broadleaf forest began again to reassemble.

Chinese vegetation, being much less subject to glaciation, is, as has been shown, a source of all manner of attractive ornamentals. At the species level, glaciation saw to it that some vanished for ever. If we set aside our regret on that point and concentrate on the genera, the return to Europe of *Ginkgo, Metasequoia, Nyssa, Liquidambar* and *Trachycarpus*, for example, might reasonably be seen merely as the restocking of an earlier widespread, but subsequently obliterated, vegetation. For an account of Chinese forest history see Menzies (1996)

Present Day Forestry Production

Since the 1980s it has been government policy to establish a remarkable total of more then 600000 ha of economic forest per year. By 1994 China possessed in total more than 160 million ha of such forest with an annual output including, 400000 tons of rosin and 150000 tonnes of paper products.

Some Chinese Forestry Trees

From northwest to southeast China the climate becomes wetter. From north to south, winter temperatures tend to rise. The consequences for the tree flora of China are that as a first approximation, one can recognise northern boreal temperate broadleaf deciduous and tropical monsoon as three major types. The first has a strong conifer content, the second is characterised by elm, oak, poplar and willow, for example, and the third has a larger collection of genera not only of trees but also of other species.

This simple threefold pattern needs to be understood from three separate viewpoints. Firstly, since the environment is not discontinuous, inter grading of these forest types can occur, secondly, this simple classification can be endlessly subdivided by taking account of soil type and aspect or other features. Thirdly, it

Table 5. 2. References relevant to forestry   (Menzies1996)

Approximate date	Location / event
Early	*Erh Ya* *The Literary Expositor* - includes references to elm and *Cunninghamia*
90 B.C.	*Shih Chi* *Historical Records* – refers to bamboo in plantations
160 A.D.	*Ssu Min Yüeh Ling* *Monthly Ordinances for Four Sorts of People.*
460	*Chu Phu* *Monograph on Bamboos* – Details of morphology etc.
535	*Chhi Min Yao Shu* Includes references to tree management
659	*Hsin, Hsiu Pên Tshao* (or *Thang Pên Tshao*) – includes a careful description of the differences between poplar and willow
1049	*Thung Phu Monograph on Paulownias*
1504	*Chu Yü Sha Fang Tsa Phu* *Miscellany from a Mountain Lodge on Bamboo Island* – includes silviculture applied to 44 species of trees and economic perennials

is necessary to appreciate that such a mosaic of tree vegetation has been radically modified by various sorts of exploitation, notably clearance and then, sometimes, replanting with species other than those native to the area. What we see in contemporary China is the outcome partly of, here and there, up to 3000 years of human activity.

Although Chinese literature is extensive, few documents, especially early on are committed exclusively to forestry. Rather, such details as do emerge are incidental to agriculture more generally. Table 5.2 includes several references, some mentioned previously in a different connection. From the literature in Table 5.2 and other historical sources there emerges a forestry technology based primarily on bamboo, *Catalpa, Paulownia, Populus, Quercus, Salix* and *Ulmus* together with the conifers, primarily but not exclusively *Cunninghamia* and *Pinus*. For a recent botanically based assessment of forestry in China see Zhang (1998)

Forestry in a Wider Context

A legend surviving from ancient Greece concerns Erysichthon who, despite the advice of the goddess Demeter, cut down a grove of trees to build a banqueting hall. This completed, he found that, when he ate, he simply became more and more hungry and eventually ended up in desperate straits as a beggar eating filth from the gutter. It is a salutary tale deserving careful reflection. By contrast, to plant trees one has to take the long view, but to do so creates all manner of benefits. The Chinese involvement in forestry is instructive.

Forestry is sufficiently important for there to be 10 forestry universities and 59 secondary specialised forestry schools. Nearly 250 institutes are committed to some aspect of forestry research. It is hardly surprising, therefore, that interest and commitment extends beyond just timber production. Three further aspects are selected for comment.

1. Environmental Protection. Across the north of China there is environmental deterioration due to desertification. To match the scale of the problem there has been a massive response – the Three North Shelterbelt scheme more widely known as Chinas Green Great Wall. The statistics are breathtaking. Across 13 provinces is a total of 4069m km^2 or 42% of China's land area. From 1978 to 2050 it is planned to plant 35.08 m ha of forest across this region, making it the world's largest ecological programme. Although on a smaller scale there are still huge afforestation programmes in progress to diminish soil erosion along the Yangtse River and to impede erosion along virtually the entire seaward boundary.

2. Animal Protection. Some 6347 species of vertebrates occur in China – approaching 14% of the world's total. Some occur elsewhere, but others are unique to China, among them endangered species such as the giant panda, golden–haired monkey and numerous threatened bird species. Since so many animals inhabit the forest, the Ministry of Forestry has been made responsible for the country's numerous nature reserves - a matter considered in more detail in chapter 15.

3. Forest Tourism. Anyone familiar with China over the past decade is aware of the growth in tourism and, enterprisingly, forestry has claimed a share of this. Following the creation of Zhangjiagie National Forest Park in 1982 there were by 1994 no fewer than 234 such parks, with room for many more. In that year

China inaugurated, in anticipation of demand, its International Forest Travel Service catering for both tourism and hunting.

From this overview of China's trees, attention moves in the next chapter to growth forms arguably derived from them. In so doing, the aim is to convey yet more of the interest and diversity of the Chinese flora.

Plate 1

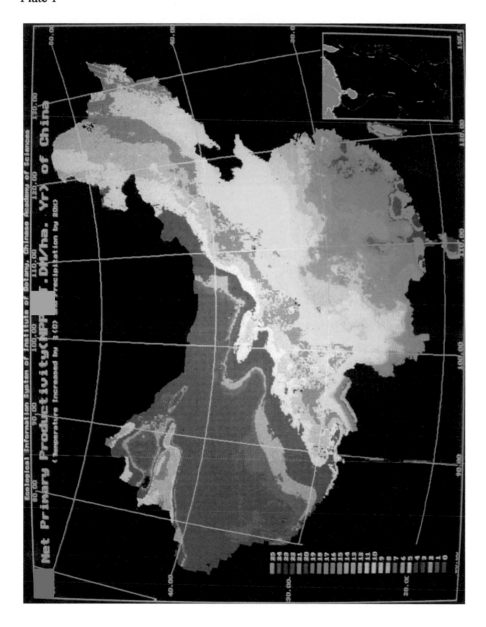

Plate 2

Plate 3

Plate 4

Plate 5

Plate 6

Plate 7

Plate 8

Plate 9

Plate 10

Plate 11

Plate 12

Plate 13

Plate 14

Plate 15

Plate 16

Plate 1

Net primary productivity. Low productivity is evident to the north and north-west but which can in some instances be raised by irrigation. Regions of low productivity can be botanically diverse with slow growing plant life distributed across a range of habitats. Higher productivity (and with it great botanical diversity) is a feature of the southern part of China especially toward the West

(Reproduced by courtesy of the Institute of Botany, Beijing)

Plate 2

(a)   Western Mountain near Kunming, Yunnan. A dissected limestone landscape with scattered conifers and a rich ground flora.
(b)   *Scaevola taccada* (Goodeniaceae) on the coast of Hong Kong.
(c)   Xinglong Mountain nature reserve, Gansu.
(d)   Great Wall of China near its crossing with the Silk Road in Gansu. Here the Wall is about 4 metres high and substantially eroded. The portion near Beijing remodelled for the Ming Emperors is not typical of its whole length.

Plate 3

(a)   Typical appearance of the Loess Plateau. In the hollow is a market garden.
(b)   Terraced loess near Lanzhou, Gansu. The banks are stabilised with cypress trees, irrigation is available and the terraces are farmed.
(c)   The panda reserve in Wolong, Sichuan showing the geological upheaval.
(d)   Rubber plantation in Hainan. The island produces, too, tea and vanilla.

Plate 4

(a)   *Acer griseum* (Aceraceae), maple. This species provides impressive autumn colour and has, too, a deciduous bark.
(b)   *Acontinum volubile* (Ranunculaceae), monkshood. A scrambling species which like all others in the genus is poisonous.
(c)   *Chimonanthus praecox* (Calycanthaceae), wintersweet. A fragrant shrub, hardy and flowering in January in the Northern Hemisphere.
(d)   *Pterocarya stenoptera* (Juglandaceae), Chinese wing nut showing its characteristic strings of fruits.

Plate 5

(a) *Caryopteris* (Verbenaceae), bluebeard. The illustration is of a garden hybrid of uncertain origin arising from among the five or so Chinese species.
(b) *Cornus controversa* (Cornaceae), growing in west Sichuan and showing its characteristically layered branches.
(c) *Paulownia fortunei* (Scrophulariaceae). One of ten species occurring in China and giving a substantial tree with striking tubular flowers.
(d) *Poncirus trifoliata* (Rutaceae). Trifoliate orange. This is a relative of orange with which it will hybridise. The plant is impressively thorny making it suitable for hedges. The fruit has a characteristic velvety skin.

Plate 6

(a) *Cassia swallensis* (Leguminosae). This specimen is growing under cover in Beijing Botanic Garden.
(b) *Malus floribunda* (Rosaceae). This popular ornamental of Western gardens is of uncertain origin possibly from China or Japan
(c) *Polygonum orientale* (Polygonaceae). This species makes a shrub perhaps 2 metres tall with showy flower racemes.
(d) *Syringa* x *villosa* (Olaeceae), lilac. Material here was part of a hybridisation programme in Beijing Botanic Garden.

Plate 7

(a) *Cephalotaxus fortunei* (Cephalotaxaceae), plum yew, cows tail pine. One of six species. Others are of medicinal interest and some are threatened through over-collection.
(b) *Abies forrestii* (Pinaceae), silver fir. A locally distributed tree in China growing between 2500 and 4200 m a.s.l. in S. W. Sichuan, E. Xizang and N. W. Yunnan. This and the following species have attractive silvery foliage and mauve/purple coloured female cones.
(c) *Abies delavayi*, silver fir. A locally distributed tree in S. E. Xizang, N. W. Yunnan and into Myanmar. As with the previous species it can provide high quality timber.
(d) *Pinus bungeana* (Pinaceae), lacebark pine. One of three pine species having an attractive flaky bark on mature trees and an appealing choice for ornamental plantings.
(e) *Pinus bungeana* showing female cone toward maturity.
(f) *Picea obovata* (Pinaceae), spruce. This species occurs on mountain slopes from 1200 to 1800 m in Xinjiang, Kazakstan, Mongolia and Russia. Resin is seen conspicuously exuding from the cone. Its status in China is vulnerable.

Plate 8

Cathaya argyrophylla (Pinaceae). Only discovered some 50 years ago this rare conifer is known to occur in Guangxi, Hunan, Sichuan and Guizhou. It is a protected plant and efforts are in hand to propagate it and establish larger plantings.

(Photograph reproduced by kind permission of Royal Botanic Gardens, Kew)

Plate 9

(a) *Paeonia* sp. (Paeoniaceae), tree peony. Until recently this would have been referred to as P. *suffruticosa* but that name is now invalid. (see text)

(b) Moutan tree peonies are distinguished partly by a membrane surrounding the caspels and which ruptures as the flower matures. Its function is uncertain but is thought to impede self-pollination. The illustration shows a triangular portion of the red membrane removed and exposing the pale green carpel surface within.

(c) *Clematis rehderiana* (Ranunculaceae Section Viorna, sub section Connatae). Native to W. China this species is scented.

(d) *Clematis tangutica* Section Meclatis. Growing in N. W. China this specimen shows flowers and fruits. The species is widely grown in temperate gardens.

Plate 10

(a) *Kolkwitzia amabilis* (Caprifoliaceae), beauty bush. Located in central and eastern China the species is now rare through land exploitation in its natural habitats. It is however readily obtainable in Western horticulture from commercial nurseries.

(b) *Musella lasiocarpa* (Musaceae). Native to Yunnan, this is apparently little known in Western horticulture although it would be an attractive subject for conservatories or subtropical/warm temperate gardens.

(c) *Tricyrtis formosana* (Liliaceae), toad lily. Several species are known of which this is the commonest and most easily grown.

(d) *Liriope muscari* (Liliaceae) turf lily. Interesting from the viewpoint of the present text in that it occurs in mainland China, Taiwan and Japan.

Plate 11

(a) *Anaphalis lactea* (Compositae). An ingredient of alpine pasture. The genus occurs in upland areas of the north temperate region.

(b)   *Meconopsis* (garden cultivar) (Papaveraceae) Himalayan blue poppy. The plant, of unknown origin, has affinities with *M. delavayi* and *M. quintuplinerva* but with flowers up to 16 cm in diameter when fully expanded.

(c)   *Gentiana dahurica* (Gentianaceae) a Chinese representative of a family occurring in temperate regions of both northern and southern hemispheres.

(d)   *Bos grunniens* (Bovidae), yak. The animals are found on alpine pasture. A first generation cross with domestic cattle can yield a pian, a handsome draught animal of considerable strength.

Plate 12

(a)   *Primula capitata*, (Primulaceae Section Capitatae). This alpine species carries numerous flowers in a flattened head. In nature it is distributed through Nepal, Xizang and Myanmar.

(b)   *Primula pulverulenta*, (section Proliferae). This species contrasts with the former in having succeeding whorls of flowers. The photographer is Yuying Geng Curator of the Huaxi sub-alpine botanic garden, Sichuan.

(c)   *Rhododendron ambiguum*, (Ericaceae Section Rhododendron, sub sect. Triflora). Native to Sichuan.

(d)   *Rhododendron auriculatum*, (Section Pontica, sub sect. Auriculata). Native to Guizhou, Hubei and Sichuan, it makes a large shrub and with fragrant flowers.

Plate 13

(a)   *Rhododendron calophytum* (Section Pontica, sub sect. Fortunea). This species native to Yunnan and Sichuan will form a tree up to 12 m tall.

(b)   *Rhododendron oreodoxa* (Section Pontica, sub sect. Fortunea). Found in western China this specimen is growing in the Huaxi collection in Sichuan.

(c)   *Rhododendron polylepis* (Section Pontica sub sect. Triflora). A handsome species native to Sichuan.

(d)   *Rhododendron racemosum* (Section Rhododendron sub sect. Scabrifolia). This is native to Yunnan and Sichuan and has been used as the parent of garden hybrids rather than a specimen plant in its own right.

Plate 14

(a)   *Rhododendron cinnabarinum* var. 'Cinzan'. The species belongs to Section Rhododendron sub sect. Cinnabarina and is sub divided into three subspecies of which two occur in China and beyond and one (*tamaense*) is confined to Myanmar.

(b) *Rhododendron hunwellianum* (Section Pontica sub sect. Argyrophylla). The species occurs in Sichuan and Gansu.

(c) *Rhododendron monosematum* (Section Pontica, sub sect. Maculifera). Normally with some colour in the corolla this species is found in Yunnan and Sichuan and in the illustration represented by a striking all–white variant

(d) Variety 'Pink Pearl'. For a discussion of its provenance see Chapter 12.

Plate 15

(a) *Rosa sericea* f. *pterocantha* (section Pimpinellifoliae, Rosaceae). Known from India, Myanmar and western China in elevated locations the form shown here has ornamental thorns. In tall, well-grown specimens the thorns can appear translucent with the sun shining through them.

(b) *Rosa banksiae* (section Banksianae). Showing here is the shaggy deciduous bark of a mature stem. The species, native to western China is widely known horticulturally in the forms *lutea* and *lutescens*.

(c) *Rosa xanthina* (syn. *hugonis*, Section Pimpinellifoliae). Occurs in western China and has generated such famous roses as Canary Bird and R x *cantabrigiensis* the latter having been derived as a hybrid in the University of Cambridge Botanic Garden.

(d) *Rosa moyesii* (Section Cinnamomeae). Originating from N. W. China, several variants of this species occur in cultivation of which the most famous is the absurdly named Geranium. The bottle-shaped hips enliven autumn and winter gardens.

Plate 16

(a) *Cunninghamia lanceolata* (Taxodiaceae), China fir. Generally considered to be a single though variable species, a fairly recent development has been the recognition of a second species *C. unicaniculata* restricted to S. W. Sichuan and regarded as endangered.

(b) *Pseudolarix amabilis* (syn. *P. kaemferi*, Pinaceae). Native to S. E. China this plant, now rare, is both an ornamental and a timber tree. Among a conifer collection this species adds interest since, along with *Larix* and *Metasequoia*, for example, it is deciduous.

(c) *Liriodendron tulipifera* (Magnoliaceae), tulip tree. The genus has one variable species which is widely distributed in eastern N. America, China and Indo China and is known elsewhere from Tertiary fossil material. Under favourable conditions it can make massive growth. One tree in upland Madeira has a trunk in excess of 7 m. in circumference.

(d) *Paphiopedilum bellatum* (Orchidaceae). The specimen shown here was seized as contraband by Hong Kong Customs and Excise officials and transferred to a secure unit at the Kadoorie Foundation for eventual relocation to the wild.

Note: Except where stated, all photographs are by G. P. C.

6 Shrubs and Climbers

The previous chapter provides some awareness of the Chinese tree flora. This chapter and the subsequent two utilise changed growth habit as an entry point into questions of botanical interest in a Chinese context.

Angiosperm Origins

The emergence of the angiosperms from whatever was their immediate ancestor could well have taken place within a mostly forest or woodland situation. If this were the case initially arborescence would have been an advantage. If one looks ahead from this, angiosperms became conspicuous in three ways, namely vegetative diversity, floral elaboration and eco-geographical spread. What, though, characterised the opening episodes of angiosperm diversification?

A Speculative View

That a malleability of sorts persists is evident, for example, in what plant breeders are able to achieve. Even so, they work within definable taxonomic limits. Boundaries to crossability do exist and this has prompted, through molecular biology, the search for means to cross such boundaries. Let us, therefore, suppose that at the dawn of the angiosperms three things co-existed. Firstly, through close relationship there was an absence of crossability barriers. Secondly, a rate of mutation existed to generate an adequate supply of genetic alternatives. Thirdly, a benignly diverse environment allowed the survival of all manner of oddities. Under these circumstances we could envisage woody plants giving rise to progeny which eventually included shrubs and climbers and, at the extremes of neoteny, thorough–going herbs. Was it perhaps only after this diversification was well under way that the crystallisation of taxonomic boundaries became significant? If not how else might we explain the range of vegetative form among the globally successful families?

A long-standing puzzle has been the relation between mono- and dicotyledons. Which, for example, came first and from what part of the other's range and type did they arise? We do not know but what *is* evident is that both groups contain not merely trees, shrubs and herbs but each has a diversity of subtypes for which

Fig. 6. 1. A Proposed Summary of dicotyledon - monocotyledon relations

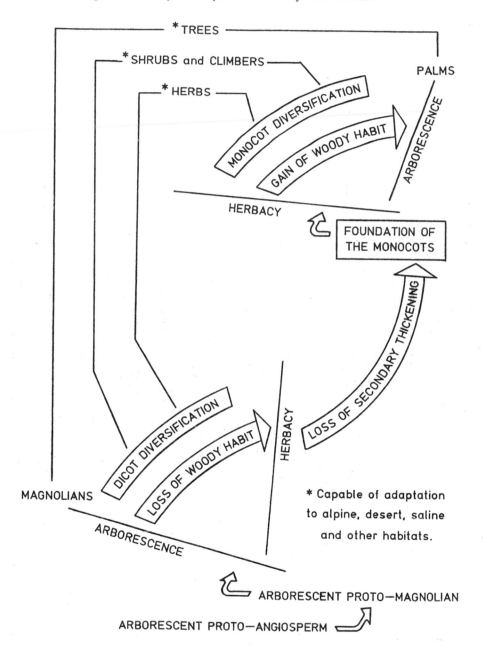

the rationale is ecological. As a backdrop, therefore, to chapters 5, 6, 7 and 8 we offer Fig 6.1

The principal assumptions of Fig 6.1 are that the immediate ancestors of flowering plants were arborescent and that among living plants today the "magnolians" include their closest surviving relatives. From these or something like them there has been both diversification and loss of woodiness, leading to thorough–going herbacy, and that out of such lineage the monocotyledons arose. These initially herbaceous types regained woodiness, of a kind, finding its most complete expression among the palms.

One especially thought–provoking aspect of this view from the magnolians through to the palms is the durability of double fertilisation, endosperm formation (if not its invariable persistence) and the closed carpel. Equally intriguing is the spasmodic appearance of those variants of the eight–nucleate embryo sac summarised by Maheshwari (1950) which occur in both mono– and dicotyledons, as does apomixis.

There is a further consideration. In such globally successful families as the Compositae, Leguminosae, Orchidaceae and Poaceae, for example, there is a harmony about their floral mechanisms, where all the stages from pollination to seed dispersal fit together so effectively. Such winning combinations, for that is, literally, what they are, place them at an advantage in a wide range of habitats.

This is not the whole story, since conversely we can find examples where a combination of features has allowed some taxon to survive under specialised circumstances and we need to ask if this is innovation waiting in the wings or the demise of a group formerly more widely spread.

If the main outlines of angiosperm radiation were established early, later evolution would be primarily increasing refinement mixed with what new opportunities opened for exploitation. As later events overlay earlier ones and adaptation moves back and forth we have to recognise that present–day flowering plants are the representatives of a convoluted history. Climatic change and geological upheaval have left their mark. The aim in choosing the examples which follow is to reveal some, at least, of the factors which have shaped the Chinese flora we see today. Dicotyledons are again subdivided here into Archichlamydeae and Metachlamydeae and these precede monocotyledons. Within each category families are presented alphabetically. Where appropriate, genera likely to be familiar are used to introduce more obscure Chinese relatives.

Dicotyledons

ARCHICHLAMYDEAE

Actinidiaceae

The family, typically climbers, occupies parts of east Asia and north Australia and includes *Actinidia chinensis*, the Chinese gooseberry or kiwi fruit, and *A. kolomicta* an ornamental with leaves coloured green, pink and white.

In China the endemic genus *Clematoclethra* occurs with one species C. *scandens* divided into four subspecies (Ying et al 1993). It is well represented in Yunnan, thinning out toward Gansu and Shaanxi.

Cornaceae

Spread across the Northern Hemisphere and on to tropical mountains, the most familiar representatives are in the genus *Cornus*, the dogwoods with typically red stems, and *C. mas*, the edible cornelian cherry. The taxonomy of the family is problematic, both as to the limits of *Cornus* and the wider relationship of the family itself. An interesting example here is the curious genus *Helwingia*. It is distinguished by umbels, which spring from the leaf (or strictly, cladode) midrib, reminiscent of the more familiar but unrelated *Ruscus* (Liliaceae). *Helwingia* is variously placed in Araliaceae, Cornaceae or even Helwingiaceae.

Helwingia is not endemic to China but is confined to what Good (1964) refers to as Sino-Japan, arguably a more realistic and cohesive phytogeographic area.

Lardizabalaceae

Peculiarly, the family is distributed in eastern Asia and Chile, raising the question as to whether it is a natural aggregation of related taxa. Of the seven genera recognised, two, *Boquila* and *Lardizabala,* are confined to Chile, the remainder being in eastern Asia. These include *Akebia,* of which *A. quinata* and *A. trifoliata* are familiar scented vines of temperate horticulture. *Sinofranchetia,* as *S. chinensis,* is a monotypic genus found in central China and is a climber, as is typical of the family.

Leguminosae (Fabaceae)

Although this family might have been considered either in the previous chapter on the basis of its trees or the next one because of its herb representatives, it is taken here, since the two endemic genera are of this type. *Craspedolobium schochii* is a monospecific genus and is a woody climber found in Yunnan and Sichuan. *Salweenia wardii* is an evergreen shrub of Sichuan and eastern Xizang.

Wisteria, commonly thought of as oriental, occurs naturally in both Sino-Japan and eastern north America. The oldest specimen in England is, allegedly, at Hampton Court, dating from perhaps the end of the 18th century. *Wisteria* is relatively adaptable around the world, and Florence in Italy, for example, already an attractive city, is enhanced in spring by numerous of these plants growing among the buildings.

A common wild shrub in Europe is *Ulex* (furze, gorse). Normally spiny, the first few leaves it produces can be trifoliate, especially under high humidity. *Caragana*, a shrub common in northern China, and, in some species, having yellow flowers sometimes shows resemblance to gorse and it is interesting here to examine the two genera together. *Caragana* exists in a range of about 60 species encompassing a greater diversity than *Ulex* with about 20. At perhaps one extreme are *C. sericea* and *C. tangutica*, leafy and in the former case lacking spines. At the other is *C. jubata*, a chamaephyte plant found in the mountains of Xizang literally draped in spines about 5–7 cm long and with small pinnate leaves concealed among them. Between these extremes, morphologically, one might place, say, *C. spinosa*. When the morphology is considered in more detail the situation is as follows.

Table 6. 1. A comparison of *Caragana* and *Ulex*

Caragana	*Ulex*
Depending on species leaves even – pinnate with 2–18 leaflets. These latter can be from 2n–50mm in length	Leaves 3–foliate on young plants. Later leaves develop as tripartite spines
In some species the axis of the pinnate leaf (rachis) is spiny–pointed and persists long after leaflet abscission as a formidably hard spine	
Stipules present sometimes but not invariably spiny or persistent	Stipules absent
Flowers commonly yellow, in a few species sometimes purple	Flowers yellow

Loranthaceae

Familiar through mistletoe, *Viscum album*, the Loranthaceae is most prolific under tropical conditions, where it occurs in a range of habitats. The plants, though green, are semi–parasites fastening on to the host plant through haustoria or modified roots. Established plants then hang from branches of their host. *Taxillus*, a genus in this family, is distributed from South Africa, through Madagascar and south China into Malaysia, suggesting a long–established history preceding much of the continental drift.

Melastomataceae

Among many of its genera the family is recognised by the 3–9 palmately veined leaves. Relatively well–known ornamentals include *Centradenia, Heterocentron, Medinilla* and *Tibouchina*, this last having strong violet– to purple–coloured flowers up to about 17 cm in diameter. Since each of these genera is frost–sensitive, they are, for temperate gardens, only useful as conservatory plants.

As might be expected, the family is tropical and subtropical. In China it has its principal distribution in the south and west. The endemic genera are shrubs or subshrubs and include *Barthea, Cyphotheca, Fordiophyton* (just reaching into N. Vietnam), *Scorpiothyrsus* and *Styrophyton*. Two others, *Stapfiophyton* and *Tigridiopalma*, are herbs.

To these seven can perhaps be added an interesting eighth case, *Sporexia*. If the genus is narrowly defined it consists of a few species, one of which extends into northern Myanmar. On this reckoning it is "almost endemic" to China. Later revision of the genus (Hansen 1990) has added five species and as such it now extends substantially into Myanmar.

Paeoniaceae

In view of its long held interest for the Chinese this genus was mentioned in Chapter 1. Any tourist looking at traditional Chinese fabric patterns will recognise peony flowers and leaves as a dominant theme. Originally considered part of Ranunculaceae, several distinctive features support the notion of a separate monogeneric family. These include possession of a floral disc, centrifugal dehiscence of stamens and large arillate seeds. Since both shrubby (tree) peonies and herbaceous perennial types occur, the genus could be treated here or subsequently. It is included here on the grounds that woody habit is presumably a surviving archaism or plesiomorph and the herbaceous habit derivative and thus apomorph.

Figure 6.2 shows a simplified classification of the genus *Paeonia*. Tree peonies are Asiatic while herbaceous types occur more widely. Emphasis, therefore, in the present discussion is given initially to the tree peonies. Traditionally referred to as *P. suffruticosa*, (Plate 9) this name is now abandoned and several others used instead. The distinction between the two subsections, Vaginatae and Delavayanae, is based on the former having a 'disc', a leathery sheath around the fruit.

In China interest in tree peonies is threefold – medicinal (see Chapter 9) horticultural and through artistic representation. As regards horticulture, particular interest attaches to floral variation in both shape and colour. About such variants here, as with *Camellia, Clematis* and *Rosa* together many other genera, one can take either a botanical or horticultural view. The former recognises as its starting point a single whorl of petals with numerous stamens. Variants occur which convert stamens to petals give rise to doubleness. Horticulturally, interest, and indeed enthusiasm, centre on the degree of doubleness and the search for forms considered superior. Nowadays all manner of such off-types are available, bearing varietal names.

Fig. 6. 2. A simplified classification of *Paeonia* (After Page, 1997: Chromosome numbers from Darlington and Wylie 1955)

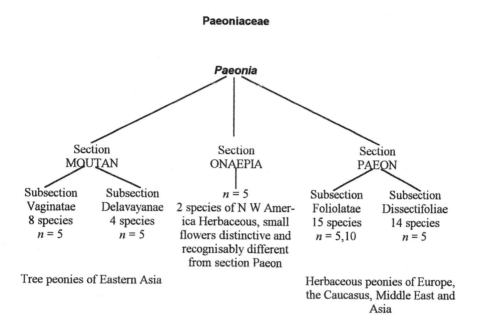

Paeoniaceae

Paeonia

Section
MOUTAN

Section
ONAEPIA

Section
PAEON

Subsection
Vaginatae
8 species
n = 5

Subsection
Delavayanae
4 species
n = 5

n = 5
2 species of N W Amer-
ica Herbaceous, small
flowers distinctive and
recognisably different
from section Paeon

Subsection
Foliolatae
15 species
n = 5,10

Subsection
Dissectifoliae
14 species
n = 5

Tree peonies of Eastern Asia

Herbaceous peonies of Europe,
the Caucasus, Middle East and
Asia

Returning to the wild tree peonies, the following species are recognised (Page, 1997) Sect. Vaginatae : *P. jishanensis* (Shanxi), *P. astii* (Shaanxi, Henan), *P. givi* (Hubei), *P. rockii* (Gansu, Sichuan, Shaanxi), *P.szechuanica* (Sichuan), *P. yananensis* (Shaanxi) and *P. yunnanensis* (Yunnan), Section Delavayanae : *P. delavayi* (Sichuan, Yunnan), *P. lutea* (Yunnan) and *P. potania* (Sichuan and Yunnan). Typically, tree peonies are upland plants occurring at around 3000 m in Yunnan, for example, and generally frost–hardy.

Perhaps the most surprising development in breeding peonies was the successful cross between the tree peony Alice Harding and the herbaceous type Kakoden. Named after their Japanese originator, such Itoh hybrids have foliage reminiscent of tree peonies but die down in winter. They are greatly esteemed among enthusiasts, (Page 1997).

One theme of the present book is the relative escape from glaciation in China compared to Europe. From classical cytogenetics, a customary assumption is that diploid species can be hybridised to yield more or less sterile progeny, which, on doubling to tetraploidy, attains improved fertility. There is also the possibility, in some ways harder to verify, that such hybridisation between diploid species can retain sufficient fertility and be perpetuated at the diploid level. To these situations, long suspected in peonies, fresh light has been brought involving nuclear ribosomal DNA. The European tetraploid *P. russi* ($2n = 20$) is shown, almost certainly, to have been a derivative of the Asiatic *P. lactiflora* and *P. mairei*. How can such wide separation be explained? Probably once indigenous to Europe, the two diploids, having originated *P. russi* (a more hardy and adaptable tetraploid), were eliminated from there during glaciation (Sang et al, 1995). Later studies include (Sang et al 1997) on chloroplast DNA, Zhang and Sang (1998) on ribosomal RNA and Sang and Zhang (1999) using the *Adh* gene.

Another area of interest is floral biology, where recently, for example, in a study of low seed set in *P. jishanensis* it was concluded that the self-incompatibility system was activated when the pollen tubes reached the ovary wall, (Zhou et al 1999).

Ranunculaceae

Typically a northern temperate family, it is the source of numerous ornamental genera of which *Aconitum*, *Clematis* and *Delphinium* have provided significant contributions. Less well known are seven endemic genera, herbaceous and considered in the following chapter. *Anemone*, together with its close congeners, *Hepatica* and *Pulsatilla*, are considered separately in Chapter 11.

Opportunity is taken here to discuss the genus *Clematis* (Plate 9). *Aconitum* has a medicinal significance and reference is made to it in Chapter 9.

Clematis contains about 250 species, of which about 100 grow in China, with characteristically thigmatropic petioles, which curl round supports with which they come in contact. Where the style persists, it provides, as a result of its hairy growth, means of wind-distributing plumed achenes.

Botanically, generalisations are useful at his point. Flowers are tetramerous, usually, that is to say they have four perianth segments. Secondly, this phrase is deliberately chosen since the flower parts are not readily divisible into petals and sepals. Since, thirdly, the stamens are numerous, the variant forms can be expressed as staminodes or extra petals (the *flora plena* or double forms). Fourthly, the floral variation, whether in perianth or stamen number, is similar in both the New and Old World species. The overall impression is, perhaps, of a long–established genus, well spread across the Northern Hemisphere and reworking similar themes throughout its range in the context of broadleaf woodland. Apart, however, from one species, *C. texensis*, only Old World species are involved in the production of commercial and usually large-flowered hybrids.

As with other genera, horticultural and botanical classifications can diverge widely, and currently the latter subdivide *Clematis* to ten or so sections perhaps thereafter dividing these to subsections and even series. Of the many species in China, the following are of some significance in temperate gardens, *C. apiifolia*, *armandii*, *connata*, *florida*, *henryi*, *heracleifolia*, *hexapetala*, *kirilowii*, *macropetala*, *montana*, *patens*, *potaninii*, *rehderiana*, *tangutica* and *uncinata*. (It should be recognised that species names can change as taxonomists revise the work of their predecessors). Part of their interest for us is the ability to throw unusual variants. For further details see Fretwell (1989). For a more botanically orientated treatment, an excellent recent addition is that of Grey-Wilson (2000)

Table 6. 2. Subdivision of the genus *Clematis* for species occurring in China – (After Wang 1980, 1998)

Section	Subsection	No. of species	Examples in horticulture
	Crispae	2	
Viorna	Tubulosae	3	*heracleifolia*
	Connatae	31	*connata, rehderiana, henryi*
Atragene	-	4	*macropetala*
Meclatis	-	8	*tangutica*
Fruticella	-	6	
Clematis	Augustifoliae	1	*hexapetala*
	Rectae	14	*armandii, kirilowii, uncinata*
	Crasifoliae	1	
	Vitalbae	15	*apiifolia*
Viticella	-	8	*florida, patens*
Cheiropsis	-	8	*montana, potaninii*
Naraveliopsis	-	7	
		108	

Rosaceae

This family is cosmopolitan, though recognisably more temperate than tropical. Familiar genera, including *Malus, Prunus, Pyrus* and *Rubus*, are commonplace in China. *Prunus triloba* is a conspicuous item here of the spring flora in street plantings and parks. Ornamental species deriving from China are common in temperate gardens. If one takes, for example, the broader canvas of Sino-Japan, the region can claim *Kerria* and *Eriobotrya*. Three endemic genera of Rosaceae occur in China, *Dichotomanthes* in Yunnan and Sichuan, *Spenceria* in Xizang and Yunnan and *Taihangia* in Henan and Hubei. The first is a shrubby perennial and the other two perennial herbs.

Chinese representatives of the genus *Rosa* are of considerable significance in commercial rose breeding and these are dealt with in Chapter 13.

Sargentodoxaceae

The family, having but one genus and two species, is almost endemic to China but just extending into northern Laos. Both species are climbers, *Sargentodoxa cuneata* being dioeceous and *S. simplicifolia* andromonoceious.

The genus is of interest medicinally having, apparently, some value in the treatment of appendicitis, irregular menstruation and arthritis.

METACHLAMYDEAE

Aristolochiaceae

The Aristolochiaceae occur in tropical and warm temperate regions and include the familiar American ornamentals *Aristolochia elegans* and *A.grandiflora*. One other species, *A. fimbriata*, has flowers strongly reminiscent of the leaf pitchers of *Nepenthes* and, given the efficiency of both in trapping insects, provides an interesting example of evolutionary convergence.

There is one genus, *Saruma*, endemic to China, and its single species, *S. henryi*, is a perennial herb found in Jiangxi province.

Caprifoliaceae

This family is common to temperate areas and the colder parts of tropical mountains. Its representatives range from small trees through shrubs and climbers to, occasionally, herbs. Important ornamentals include *Abelia, Diervilla, Leycesteria, Lonicera, Symphoricarpus* and *Viburnum*, and to most of these there is a contribution from Chinese species. That said, they are probably all outshone by *Kolkwitzia amabilis*, (Plate 10) a monotypic genus confined to China. This, the Beauty Bush genuinely deserves the adulation accorded it. Two other endemics in this family are *Dipelta*, a deciduous shrub of some horticultural interest through *D. floribunda* and *D. yunnanensis* and through the other genus *Heptacodium*. The three genera are spread across southern China.

Ericaceae

Phytogeographically, and by relationship, the family can be subdivided into an ericoid group common in Africa, the Mediterranean margin and Europe, a vaccinoid group typically American and Asiatic and a rhododendroid group found in Eastern Asia and North America. This is not to say that distributions do not overlap because clearly they do, as with both *Erica* and *Vaccinium* indigenous to the European flora, for example. The distinctions drawn are, however, useful generalisations. *Rhododendron*, an ornamental of major importance, is treated separately in Chapter 12.

Other genera of note as ornamentals are *Enkianthus* from China, Japan and the Himalayas and *Pieris*, of which *P. floribunda* and *P. japonica* are representatives of American and Asiatic parts of the vaccinoid group.

Oleaceae

Although the olive, *Olea*, readily indicates a Mediterranean aspect to this family, in fact it ranges widely, especially through temperate and even tropical Asia. This is readily recognised through the contribution to horticulture, for example, of *Jasminum, Osmanthus* and *Syringa* as indicated earlier in this book. *Osmanthus*, before the effect of human travel, was entirely eastern. *Forsythia* and *Syringa* range further west into Europe, *Jasminium* was among the earliest introductions from the East to the West, dating from about the mid-16th century.

Rubiaceae

The Rubiaceae range across most latitudes and include a wide diversity of growth habit from trees such as *Genipa* through a range of shrubs and climbers to slender herbs like *Asperula* and *Gallum*. The most economically important genus is *Coffea*, native to east tropical Africa but grown more intensively in the New World. Among ornamental shrubs and climbers are *Cephalanthus, Gardenia, Ixora* and *Manettia*. Smaller plants include *Pentas*, suitable for bedding out in warmer countries. *Cinchona*, Jesuits bark provides quinine. Given the spread of Rubiaceae and its range of growth habits, it is hardly a surprise that it is so significant an ingredient of the Chinese flora although none of the genera so far mentioned here is especially significant or even indigenous to China. The family is included in this chapter because of its four Chinese endemics; one, *Emmenopterys* (southern China), is a tree, and two are shrubby. These are *Dunnia* (Guangdong) and *Trailliaedoxa* (Yunnan). The fourth, a herb, *Hayataella* is restricted to Taiwan.

Monocotyledons

Palms, the conspicuously woody monocotyledons were mentioned previously. *Cordyline* and *Dracaena* occur in southern China.

Of some interest is the range of opinion among taxonomists as to how inclusive the Liliaceae is seen to be. In the present situation, Asparaginaceae and Smilacaceae are treated here as separate families.

Asparaginaceae

Asparagus officinalis, familiar as a spring vegetable and probably originating from the Mediterranean margin, shares with its related genus *Ruscus* the production of cladodes or leaves bearing flowers – a feature already mentioned for *Helwingia* (Cornaceae). The Chinese plant *A. lucidus* produces edible tubers.

Poaceae

Grasses tend normally to be herbaceous although, in some cases, older stem material can be tough and indigestible for herbivores. Matters go further in *Phragmites*, which is excellent for thatching. Two tall grasses, *Arundo* and *Saccharum*, give rise to thick, relatively tough stems in which there is appreciable fibre. The situation goes to the extreme in bamboos, where stems are strong enough to provide material for furniture and even scaffolding.

Although in common perception associated with China, bamboos in fact occur throughout the tropics in both the New and Old World, though to a lesser extent in Africa. Nonetheless, bamboos are common in southern China. In terms of diversity, bamboos reach their climax in Yunnan. In the Xishuangbanna region there is an astonishing concentration of 13 genera containing in total some 94 species (Wang 1990). An aspect of absorbing interest here is the extent to which various ethnic groups have developed particular approaches to these bamboos. See, for example, the paper quoted above and Wang et al (1993). Although endemism does occur in Yunnan, it is known for bamboos elsewhere in China to a significant extent and is summarised in Table 6.3. so as to permit comparison with other taxa mentioned elsewhere in this book.

Table 6. 3. Distribution of endemic bamboos in China.

GENUS / Species	Yunnan	Guangxi	Hainan	Guangdong	Guizhou	Hunan	Jiangxi	Zhejiang	Sichuan	Hubei	Fujian	Anhui	Jiangsu	Taiwan
Acidosasa 6	+			+		+	+				+			
Ampelocalamus 1			+											
Bashania 4	+				+	+			+					
Brachystachyun 1								+					+	
Chikusichloa[a] 2		+	+									+	+	
Ferrocalamus 2	+													
Gelidocalamus[b] 11		+		+	+	+	+	+						
Leptocana 1	+													
Metasasa 1				+										
Monocladus 4		+	+	+										
Oligostachyum[b] 14		+	+	+			+	+			+	+		
Qiongzhuea 3	+				+					+	+			
	5	4	4	5	3	3	3	3	2	2	2	1	2	1

The table indicates fewer endemics in the north and east

[a]   One species reaches Japan

[b]   *Gelidocalamus* and *Oligostachyum* having more species, unsurprisingly are more widely spread.

To facilitate province–by–province comparison, the bold vertical lines here are placed similarly in Tables 7.2 and 7.3

For a comprehensive treatment of bamboo biology including ecology, taxonomy, flowering behaviour, evolution and anatomical structure see Chapman

(1997). This text deals, too, with the interaction between bamboos and the giant panda.

Smilacaceae

The genus *Smilax*, normally dioecious, exists worldwide in the tropics and subtropics and is well represented in China. As an example of this, in the relatively confined area of Long chi in western Sichuan there are 16 species recorded together with one *Heterosmilax*. *Pseudosmilax* comprises two species and is confined to Taiwan. Tubers of *S. china* are used to make a decoction against gout. It is also a minor ornamental in warmer temperate gardens.

From the primarily shrubby and climbing representatives considered, attention now shifts to chiefly herbaceous plants. It should, of course, be recognised that, given China's vast flora, the aim there, as in this Chapter, is to provide an informative sample so as to highlight various issues.

7 Herbs

Herbs survive in forest clearings, woodland margins and in altogether more open situations. Across a transect of progressively lower rainfall regions herbs would, if perennial, tend to be more xerophytic or otherwise, toward the extremes, escape drought by short generations evolved to exploit wet seasons. Survival to extreme cold is considered in the next chapter. Emphasis presently is upon largely mesophytic plants. As in the previous chapter, families are selected to indicate items of particular interest in a Chinese context and, where appropriate, familiar taxa are used to introduce more obscure ones.

Dicotyledons

ARCHICHLAMYDEAE

Apiaceae *(Umbelliferae)*

This well–recognised family, largely north–temperate, provides the ornamental *Eryngium* and a range of culinary herbs including *Apium, Dancus, Foeniculum* and *Pastinaca*.

Conium is the source of hemlock poison used in ancient Greece to kill Socrates[1]. The plant produces various alkaloids, notably coniine, methyl coniine, coniceine and conhydrine. Such is their toxicity that, even in dilute form, they have, apparently, no medicinal value.

As regards the Chinese flora, a feature of interest is the high number of endemics, 18 in all (Ying et al 1993). It may be that this is an overestimate since some evidence shows the genera *Chamaesium, Cortiella, Haplosphaera* and *Sinodielsia* do have representations beyond the Chinese border. None of the 18 is of horticultural significance although the roots of one of them, *Changium*, are used medicinally.

---

[1] The event is described in detail in Plato's *Phaedo*

Ranunculaceae

The family was considered in the previous chapter, but its herbs include some most noteworthy examples. *Aconitum, Anemone, Aquilegia, Caltha, Delphinium, Eranthis, Helleborus, Thalictrum* and *Trollius* are just some of those that could be chosen to underline its horticultural significance.

Certain genera are endemic to China, notably *Anomoclema, Kingdonia, Metanemone* and *Urophysa*, while *Asteropyrum, Beesia* and *Calathodes* are nearly so. Among the most unusual are *Kingdonia uniflora* (named after Kingdon–Ward's mother) and a plant more widely spread both in and beyond China, *Circeaster agrestis*. So peculiar are these two genera (each with a single species) that separate families Kingdoniaceae and Circeasteraceae have been proposed for them. For a recent view based on nuclear ribosomal DNA advocating Circeasteraceae, to incorporate *Circeaster* and *Kingdonia* see Oxelman and Lidén (1995). Since the two genera have appreciable characters in common, especially so in regard to their peculiar leaf venation, they might at least be thought to share a family. Given floral features, particularly, it seems not unreasonable to include them within Ranunculaceae, the view adopted here. Even so, their respective floral structures do differ as is evident from the following comparisons.

Table 7. 1. A comparison of *Circeaster* and *Kingdonia*

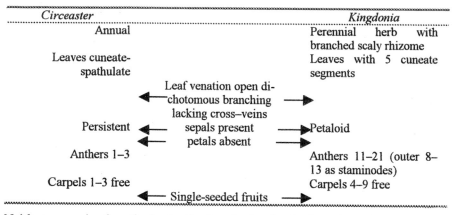

Circeaster		*Kingdonia*
Annual		Perennial herb with branched scaly rhizome
Leaves cuneate-spathulate		Leaves with 5 cuneate segments
←	Leaf venation open dichotomous branching lacking cross–veins	→
Persistent ←	sepals present	→ Petaloid
←	petals absent	→
Anthers 1–3		Anthers 11–21 (outer 8–13 as staminodes)
Carpels 1–3 free		Carpels 4–9 free
←	Single-seeded fruits	→

Neither genus is of particular horticultural value but both are botanical curiosities with most interest centred on the leaf structures. These could each be regarded initially as either a primitive survival or derived by reduction from more complex structures. It is necessary first to establish whether or not the apparent similarity between their leaf structures is genuine and thereafter to offer some view as to evolutionary status. Early studies were those of Junell (1931) and Foster (1963). More recently, Ren and Hu (1996, 1998) and Ren et al (1997) examined *Circeaster* and Ren and Hu (1996, 1998 and Ren *et al* 1998–2 refs.) examined *Kingdonia*. A separate enquiry into karyomorphology and relationships of *Circeaster* is that of Kong and Yang (1997).

The essentials of leaf anatomy are as follows. In each genus the principal vein entering the leaf base subdivides dichotomously repeatedly and the resulting veinlets, at their extremes, enter the teeth at the leaf margin. There is, unlike that found in any other dicotyledon or monocotyledon, an absence of cross–veins forming the filigree pattern familiar in partially decayed autumn leaves. Nonetheless, there are occasional instances of veins which, having divided, merge as anastomoses.

The principle clarification of leaf vasculature in these two genera was through the elegant researches of Foster. For *Kingdonia* Foster and Arnott (1960) showed how an even number of vascular bundles in the petiole supplied an odd number of laminar segments and dichotomised thereafter. Anastomoses were found to be a rare occurrence. Later, Foster (1962), studying leaf ontogeny in this genus, showed that vein branches end in or near the laminar teeth or, blindly some distance from the margin. Additionally, there was found to be no differentiation into palisade and spongy parenchyma in the leaf. Subsequently, Foster, (1966) distinguished in *Circeaster*, where anastomoses are more common, three forms of them illustrated diagramatically as follows

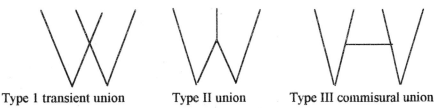

Type 1 transient union   Type II union   Type III commisural union

Such dichotomous venation seems primitive and has been compared to that in the fern *Aneimia* and the gymnosperm *Ginkgo*. Such a reversion in an angiosperm would represent an immense phylogenetic leap and, unsurprisingly, some other explanation in terms of later secondary adaptation has been sought. Later investigators, notably Ren and coworkers have extended the range of observations and reworked the arguments for these two curious genera, but the puzzles remain and admit of no clear explanation.

METACHLAMYDEAE

Asteraceae (Compositae)

The family, of global significance, deserves mention primarily because it is the largest in China. Among these are 17 endemic genera, *Ajaniopsis, Bolocephalus, Diceroclados, Diplazoptilon, Formania, Heteroplexus, Myrinopsis, Nannoglottis, Notoseris, Nouelia, Opisthopappus, Sheareria, Sinacalis, Sinoleontopodium, Stilpnolepis, Syncalathium* and *Xanthopappus* Ying et al (1993).

Table 7.2. Distribution of endemic genera of Gesneriaceae

Genus	Species	Xizang	Yunnan	Guangxi	Hainan	Guangdong	Guizhou	Hunan	Jiangxi	Zhejiang	Sichuan	Hubei	Fujian	Anhui	Jiangsu	Taiwan	Gansu	Shaanxi
Allocheilos	1						+											
Allostigua	1			+														
Aneylostemon	11		+				+				+	+					+	+
Bournea	2					+							+					
Briggsiopsis	1		+								+							
Calcareotoea	1		+	+														
Cathayanthe	1				+													
Chiritopsis	8		+			+								+				
Dayaoshania	1		+															
Deinocheilos	2								+		+							
Didymostigma	1					+							+					
Dolicholoma	1		+															
Gyrocheilos	4		+			+												
Gyrogye	1		+															
Hemiboeopsis	1		+															
Isometrum	12										+						+	+
Lagarosolen	1		+															
Metabriggsia	2		+															
Metapetrocosmea	1				+													
Petrocodon	1		+			+	+	+				+						
Primulina	1					+												
Pseudochirita	1		+															
Rhabdothamopsis	1	+					+				+							
Schistolobos	1		+															
Tengia	1						+											
Thamnocharis	1						+											
Tremacron	7	+									+							
Whytochia	3	+	+				+	+				+				+		
		8	12	2	6	7	2	1	0		5	4	2	1	0	1	2	2

Note: 1 Genus - *Hemiboea* is omitted here since it is widely distributed throughout southern China from Xizang to Taiwan.
For comparison see Tables 7.1 and 8.2

While, as the reader might have come to expect, some of these endemics are found in Sichuan, Xizang and Yunnan, others are more widely spread not only across southern China but also toward the drier, more open, habitats of the north.

Table 7. 3. Distribution of endemic genera of Lamiaceae

GENUS Species	Yunnan	Guangxi	Hainan	Guangdong	Guizhou	Hunan	Jiangxi	Zhejiang	Sichuan	Hubei	Fujian	Anhui	Jiangsu	Taiwan	Gansu	Shaanxi	Hebei	Henan
Bostrychanthera 1		+			+				+		+			+				
Cardioteucris 1	+								+									
Hanceola 6	+	+			+	+	+	+	+									
Heterolamium 1	+					+			+	+						+		
Holocheila 1	+																	
Kinostemon 2		+			+				+	+						+		
Loxocalyx 2	+					+	+		+	+					+	+	+	+
Ombrocharis 1						+												
Rostrinucula 2	+	+			+	+			+	+					+	+		
Schnabelia 2		+		+	+		+	+	+		+							
Skapanthus 1	+								+									
Suzukia[a] 2														+				
Wenchengia 1			+															
	7	5	1	1	5	5	3	2	9	4	2	0	0	2	2	4	1	1

[a] One species reaches Ryukyu Is., Japan
For comparison see Tables 6.3 and 7.2

None of the endemics has any horticultural significance, nor do they apparently have any recorded medicinal properties.

Among the genera which range beyond China, several are important in horticulture including *Aster, Chrysanthemum, Inula* and *Ligularia*. It is worth pointing out that most ornamentals of any consequence apart, from *Chrysanthemum*, in this family derive either from the New World or southern Africa. *Helianthus*, the source of sunflower oil, is primarily a North American genus.

Gesneriaceae

This is a large tropical and subtropical family found in both the New and Old World, in the Far East a conspicuous component of the vegetation. Among the surprises, therefore, are that *Haberlea* and *Ramonda*, for example, comprise alpine representatives found in Europe. Gesneriads are even more conspicuous among Chinese alpine plants – a topic considered in the following chapter. What is of special interest here is that of all plant families in China, the Gesneriaceae provide the largest number of endemics, some 29 in all. One genus, *Hemiboea,* among these reaches into northern Vietnam and the Ryukyu islands of Japan. It serves to underline the point that political and phytogeographical boundaries need not and often, do not, co-incide. Table 7.2 is set out so as to facilitate, on a province by province basis comparison with the bamboos (table 6.3) and Lamiaceae (Table 7.3).

Lamiaceae (Labiatae)

The Lamiaceae centre on the Mediterranean region but are nonetheless well represented in both the New and Old World and at temperate and tropical latitudes. The familiar genera of horticulture *Coleus, Phlomis, Rosemarinus, Salvia* and *Stachys* owe little, if anything, to China, or even so large a genus as *Salvia* with about 700 species. There are a few exceptions such as in *Scutellaria, Dracocephalum* and *Elsholtzia*. It will, however, be is evident from the genera and species present, the Lamiaceae comprise a significant component of the Chinese flora.

Among the endemics, some 13 genera are recorded, Ying et al. (1993) and these are set out by province in Table 7.3. As with other genera, the representation in Yunnan and Sichuan is noteworthy. With the exception of *Schnabelia*, used as a febrifuge, for example, none of these endemics appears to have significant medicinal application.

Monocotyledons

Liliaceae

While numerous genera from China in this family have been imported to the West, two genera only are singled out for comment here. These are *Liriope* and *Tricyrtis* (Plate 10) and are interesting because they are spread from China, through Taiwan to Japan. They provide two of numerous instances showing links between the floras of China and Japan. The Sea of Japan separating them is thought to have opened up some 17 million years ago.

Musaceae

Among monocotyledons, the orchids will, perhaps generally, be considered the most bizarre. On closer acquaintance among the monocotyledons the Musaceae must surely seem remarkable. Part of this must influence how we define its limits. At its broadest it is as follows

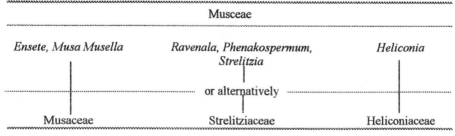

From the viewpoint of vegetative morphology it is very difficult to imagine this group as other than monophyletic. Flower and inflorescence structure are more

perplexing. By comparison with the other genera, *Musa*, for example, has mostly cream–coloured flowers subtended by bracts which, depending on species, can range from dull purple brown to the flashiest scarlet. At the other extreme *Heliconia* with, usually, a more attenuated inflorescence shows a range of bract colours from green through orange to red. *Strelitzia* differs from these in that the investment in bright colour centres on the perianth parts, typically bright blue and orange, rather than the bracts. Although among the various genera the colour and proportions of the perianth can vary, as just indicated, they are all essentially trimerous. The case for and against wider circumscription of the family is considered by Dahlgren et al. (1985) and these workers, interestingly, ascribe the bright colours of bracts (in *Heliconia*) and perianth (in *Strelitzia*) to convergent evolution, both seen as highly significant in attracting animal pollen vectors. One other genus, *Orchidantha*, sometimes attributed to Musaceae, is discussed in Chapter 15.

In *Musa*, the plant is commonly described as a giant herb and what passes for its stem is really derived from compressed, concentric leaf bases. The internal compression is readily demonstrated when the plant is decapitated and the inner leaf bases telescope upwards to varying extents, the effect being more evident toward the centre. Aside from this observation, the true stem rises through the central circle of leaf bases to emerge into and between the crown of leaves and nearing the inflorescence – initially female, briefly hermaphrodite and finally in male phase as successive bracts open to reveal their contents.

With these ideas in mind it is therefore something of a surprise to see, for the first time, the Chinese endemic *Musella* (Plate 10) found originally in Yunnan. In flower it looks like a banana plant decapitated about 60 cm above the ground and with the inflorescence prematurely emerged.

In *Ravenala* and in *Strelitzia*, for some species, a woody stem develops reminiscent of certain palms and for this reason, the Musaceae in the broad sense might have been included in Chapter 5. Given the special interest of the single species of *Musella*, *M. lasiocarpa*, and the generally herbaceous habit elsewhere in the family, it is included here.

Orchidaceae

A useful introduction to Chinese orchids is that of Yang et al. (1993). In line with expectation, Yunnan is conspicuously rich in orchids with more than 530 species representing almost 100 genera. (For comparison, the British Isles has about 50 species in 23 genera. Whereas in China, in the warmer parts, both terrestrial and epiphytic species occur, in the British Isles all orchids are exclusively terrestrial.)

While the Chinese orchid flora contains its share of both the spectacular and the humdrum, there are few, if indeed any, themes in orchid biology unique to China. That is not to say that various rarities do not excite a determined acquisitiveness among certain collectors. The matter is referred to again in Chapter 15 (See Plate 16).

Among the terrestrial orchids are a group of saprophytes devoid of chlorophyll, which include *Epigonium aphyllum*, *Galeola lindleyana* and *Gastrodia elata*, this last a protected plant. Among other terrestrial orchids selected for some degree of protection are *Paphiopedilum amenianum*, *P. micranthum*, *Changuienia amoena* and *Tangtsinia nanchuanica*. Protected epiphytes include *Bulleyia yunnanensis* and *Pleione forrestii*.

There is no doubt that, unchecked, the threat to the orchid flora of China will increase, and for more than one reason. The first is obvious, namely, the destruction of habitats through the increased human population. Significantly, too, a proportion of orchid species, sometimes involving the whole plant, is used medicinally. Thirdly, with the opening of China to tourism this will undoubtedly encourage the urge to collect. This latter tends to feed on itself in that the rarer a plant becomes the greater is the "need" in some individuals to possess it.

It is, surely, becoming apparent that collection from the wild must be regarded as increasingly old–fashioned and out of touch with the imperatives of conservation. Given the ready availability of meristem culture for many orchid taxa, vast diversity is now available and at only modest cost to the average enthusiast. At a more practical level, if the demand for folk medicines from orchids is sustained, then presumably meristem culture is inevitable for this reason alone. Relevant here is the observation that whole plants of *Dendrobium nobile* are said to be used to make either medicine or insecticide.

We now turn to an especially harsh environment, that of high mountains, but one where, again, the angiosperms demonstrate yet more of their dazzling adaptability. One surprising question is to what extent a predominantly tropical/subtropical distribution mean that such a group will not produce alpine representatives.

8 Alpines

Anyone who has climbed among mountains is aware of how awesome nature can be. Amid the quiet of great peaks one can see readily the torn and convoluted rock strata – testimony to cataclysm. Occasionally, the energy is more immediately evident in earth tremors or when volcanic eruptions burst forth. For the most part, though, what mountains reveal is the tumult of the past. At least that is the obvious conclusion to draw; but the continental plates are still in motion, the forces which shape landscape remain at work and what is written in the rocks can be a guide not only to the past but the future. Indeed, among the mountains, one has only to see one pebble fall upon another and set several more moving to realise how perched, how unstable is so much of that landscape. To appreciate more fully the character of an alpine flora, one needs some awareness of the slow but colossal dynamism of the environment within which it has evolved.

In the previous three chapters a selection of families was made to explore some of the issues which confront the plant scientist. This chapter provides a contrast in that many families and numerous genera find mention in Table 8.1 and some are selected for detailed comment. In a far larger book there would be opportunity to explore a wide range of habitats but here, the alpine environment is provided rather as a sample.

The families dealt with earlier were those with tropical, temperate and cosmopolitan distribution. Given the cold severity of high–mountain habitats, it is hardly remarkable that tropical families such as Dipterocarpaceae and Musaceae yield no alpines. More surprising is Gesneriaceae, primarily tropical, but with among its genera *Ramonda* on European mountains and, for example, *Ancyclostomon* (a Chinese endemic), *Isometrum Petracosmea* and *Rehmannia* among about a dozen found here.

It is not surprising that recognisably north temperate families such as Cruciferae, Diapensiaceae, Papaveraceae, Polygonaceae, Primulaceae, Ranunculaceae and Saxifragaceae make an appreciable alpine contribution, but it is essential to appreciate that there are few, if any, exclusively alpine families. Indeed, one can go further and point out that only a small minority of genera is exclusively montane. The common situation, by far, is that genera such as *Senecio, Cheiranthus, Euphorbia, Gentiana* and *Polygonum*, for example, have among their retinue of species a proportion of alpines, but others of this or that genus can exist in more mesophytic situations. The situation explored in *Primula* in Chapter 13 is that where the genus has an alpine quota but, also, each recognisably different parts of the genus has its subsidiary alpine subquota. The situation is more readily demonstrated in a genus as large as *Primula*, but is detectable elsewhere.

Not surprisingly, the focus of interest in this chapter is the Himalaya – Tibet massif but it is not the only alpine area. In drier northern China are regions high enough to host alpine plants. (See Plate 11)

Alpine plant enthusiasts are usually selective about what they will grow and are drawn to brightly coloured flowers and extreme modifications of growth habit. Consequently, three families which provide perhaps the bulk of alpine vegetation are consistently overlooked - Cyperaceae (sedges), Juncaceae (rushes) and Poaceae (grasses). Yet *Kobresia*, a sedge, together with *Poa* and *Stipa*, provide immense areas of alpine pasture home, for example, to herds of yak. Nor is the biology of such plants unremarkable. The reproductive versatility of *Poa,* for example, is more diverse than for any other grass of which we know and is probably one of the most remarkable among flowering plants generally.

As will be apparent from the foregoing, Table 8.1 provides a list of plants represented on Chinese mountains. While in no sense complete it does enable the following points to be made.

1. No family is exclusive to China but has there a part of its wider representation.
2. Most families are either north–temperate or cosmopolitan but can have minor representation at lower latitudes.
3. Some families are surprising in being primarily tropical in the bulk of their distribution but have evolved alpine representatives.
4. Among the genera only a very small proportion are endemic to China. At the generic level China has many in common with other mountain floras.
5. Genera indicated in bold type with large numbers of species world wide perhaps almost inevitably are well represented among Chinese alpines.

It is at the species level that endemism is rife among Chinese alpines. We can explain this partly by the creation of new habitats rather than by survival away from the extremities of Ice Ages, as tends to be the case, for example, for much of the tree flora. What follows is a consideration of several key genera, which provide instructive contrasts. For an introduction, generously illustrated, to Chinese alpine plants see Lang et al (1997)

Chamaephytic Growth Habit

The previous three chapters centered on trees, shrubs and climbers and herbs respectively. It is evident that each of these growth habits can, under appropriate circumstances, generate descendents which have become chamaephytes. *Sophora* (Leguminosae) is familiar through its tree representations but *S. moorcroftiana* is a small statured Tibetan endemic. In the same family *Caragana*, found mostly as shrub, produces in Tibet *C. gerardiana* – a short-statured prostrate plant. Examples are readily found of herbaceous genera with chemaephytic representatives. *Thlaspi andersonii* (Cruciferae) and *Aster saluensis* (Compositae), for example, are compact representatives of commonly taller herbs. It is by no means true that alpines, even those in open meadow or even scree situations are invariably compact. Alpine forms of *Rheum* produce spectacular inflorescences 1m or so long, as in *R. nobile*. As might be expected, many of these compact plants either hold their flowers well above the vegetative part of the plant, or the flowers are brightly coloured, or these two attributes are combined.

Table 8. 1. Families and genera with alpine representatives in China

(After Beckett and Grey-Wilson 1993, 1994 and augmented)

Family	Designation	Significant genera
Acanthaceae	Trop/sub trop.	*Pteracanthus*
Araceae	Trop/temp.	***Arisaema***, *Pinellia*
Berberidaceae	Trop/temp.	*Dysosma*[a], *Nandina*,
Bignoniaceae	Trop/temp.	*Incarvillea*
Boraginaceae	Trop/Temp/Medit	*Chionocaris*, ***Onosma***
Campanu-laceae	Cosmop.	*Codonopsis, Cyananthus, Platycodon*
Caprifoliaceae	Mostly temp.	***Lonicera***
Caryophyl-laceae	Cosmop.	***Arenaria***, ***Stellaria***, *Thylacospermum*
Compositae	Cosmop.	***Anaphalis***, ***Aster***, ***Cirsium***, *Cremanthodium*, ***Erigeron***, *Leontopodium*, ***Saussurea***, ***Senecio***, *Soroseris*
Crassulaceae	Cosmop.	*Rhodiola*, ***Sedum***
Cruciferae	Mostly temp.	*Cheiranthus, Coelonema,* ***Draba***, *Dipoma, Hemilophia, Neomartinella, Lignariella, Pegaephyton, Phaeonychium, Platycraspedum*[a], *Thlaspi, Solms-Laubachia*[a],
Cyperaceae	Global	***Carex***, *Kobresia*
Diapensiaceae	Cool temp.	*Diapensia, Shortia*
Ephedraceae	N & S warm temp.	*Ephedra*
Ericaceae	Almost cosmop.	*Cassiope, Diplarche, Enkianthus,* ***Rhododendron***
Euphorbiaceae	Cosmop.	***Euphorbia***
Gentianaceae	Cosmop.	***Gentiana***
Gesneriaceae	Trop/sub trop.	*Ancyclostemon*[a], *Boea, Briggsia, Corallodiscus, Isometrum*[a], *Loxostigma, Lysionotus, Petrocosmea, Rehmannia, Tremacron*[a]
Guttiferae	Mostly trop.	***Hypericum***
Hydran-geaceae	Temp/subtrop.	*Deinanthe*
Hydrocharita-ceae	Trop/temp.	*Ottelia*
Iridaceae	Trop/temp.	***Iris***
Juncaceae	Temp/trop mts.	***Juncus***, *Luzula*

Labiatae	Cosmop/Medit.	*Ajuga, Eriophyton,* **Micromeria, Phlomis, Scutellaria**
Leguminosae	Cosmop.	**Astragalus,** *Caragana, Chesneya, Gueldenstaedtia,* **Oxytropis,** *Parochetus, Sophora, Thermopsis*
Liliaceae	Cosmop.	*Clintonia, Colchicum, Fritillaria, Heleniopsis, Hosta, Lilium, Liriope, Nomocharis, Notholirion, Ophiopogon, Paris, Polygonatum, Reineckia, Rohdea, Streptopus, Theropogon, Tricyrtis*
Oleaceae	Cosmop/Asia.	*Forsythia, Syringa*
Orchidaceae	Cosmop.	**Calanthe,** *Cypripedium, Oreorchis, Pleione, Porolabium*[a], **Satyrium,** *Smithorchis*[a]
Paeoniaceae	N Temp.	*Paeonia*
Papaveraceae	N. Temp.	*Dicranostigma, Eomecon*[a]*, Meconopsis, Stylophorum*
Plumbaginaceae	Cosmop.	*Ceratostigma*
Poaceae	Global	**Poa, Stipa,** *Thamnocalamus*
Polygalaceae	almost Cosmop.	**Polygala**
Polygonaceae	mostly N temp.	**Polygonum,** *Rheum*
Primulaceae	N temp/Cosmop.	**Androsace,** *Bryocarpum,* **Lysimachia,** *Omphalogramma, Pomatosace*[a]*,* **Primula**
Ranunculaceae	mostly N. temp.	**Aconitum, Anemone,** *Caltha,* **Clematis, Delphinium,** *Oxygraphis, Paraquilegia, Semiaquilegia, Trollius,*
Rosaceae	Cosmop.	*Cotoneaster,* **Potentilla, Sorbus,** *Spenceria*[a]*,*
Saururaceae	E. Asia/N. Amer.	*Houttuynia*
Saxifragaceae	mostly N. temp.	*Chrysosplenium, Mukdenia, Oresitrophe*[a]*,* **Saxifraga,** *Tanakea*
Scrophulariaceae	Cosmop.	*Lagotis, Lancea, Oreosolen,* **Pedicularis,** *Pterygiella*[a]*, Scrofella*[a]*, Xizangia*
Solanaceae	Trop/temp.	*Atropanthe Mandragora, Przwalskia*[a]*,*
Thymelaceae	Temp/trop.	*Stellara*
Umbelliferae	chiefly N. temp.	*Pleurospermum*
Valerianaceae	Temp/sub trop.	*Nardostachys, Patrinia*
Violaceae	Cosmop.	**Viola**
Zingiberaceae	Trop/Indo mal.	*Pyrgophyllum, Roscoea*

Genera in **bold** type have 100 or more species on a world basis
[a] Genus endemic to China (After Ying et al 1993).

Selected Families

Among the gymnosperms it is possible to discern, apart from the monotypic *Ginkgo* two relatively homogenous groups, the cycads and the conifers. A third group gathered together traditionally as the Gnetales, includes the three disparate genera *Ephedra, Gnetum* and *Welwitschia*. Each has family status in its own right and 'Gnetales' is really only a convenient umbrella under which to put them. Their real interest for many botanists is to what extent, if at all, they can shed light on the origin of the angiosperms.

Ephedraceae

Ephedra, a monotypic family, is the most widely dispersed of the Gnetales with about 40 species. Several Chinese alpines are known, including *E. gerardiana, likiangensis, minima, minuta, saxatilis* and *sinica*. The genus is far older than the angiosperms, and is known in Chinese palaeobotany from the Jurassic about 150 million years ago. For ephedrine see the following chapter.

Dicotyledons

ARCHICHLAMYDEAE

Berberidaceae

The family occurs in both Northern and Southern hemispheres, often, but not invariably, in upland habitats especially in the tropics. Among Chinese representatives is the endemic genus *Dysosma* found in Tibet at up to 3500 m

Cruciferae (Brassicaceae)

Recognisably a north-temperate family, most genera are herbaceous. Of the ten endemic genera in China, eight are plants of montane areas. Species of *Draba* are particularly sought after by alpine plant enthusiasts and the genus is also well represented, on European mountains. Most *Draba* in alpine collections originate from Europe and the New World, however.

Of Chinese interest particularly are the endemic genera *Platycraspedum* with one species, *P. tibeticum* and *Solms-laubachia* with five species, which these can be found up to 5700 m in south–central and west–central China.

Papaveraceae

A family of 26 genera and about 300 species it is generally north temperate. There are 12 genera and perhaps 70 species in China of which one genus with one species, *Eomecon hionantha*, can occur as an upland endemic in forest or under shade. In an alpine context the family yields in China some impressive examples cherished by collectors around the world, of which *Meconopsis* is perhaps the most spectacular with its range of flower colours especially the electric blue species notably *M. horridula* found at about 4000 m in Tibet, Yunnan, Sichuan, Gansu and Qinghai. For a garden form of blue *Meconopsis* see Plate 11.

Rosaceae

As was evident from previous sections, this family is a significant ingredient of the Chinese flora. The four genera listed in Table 8.1 are merely a sample of alpine representatives. Particular interest attaches to *Spenceria*, an endemic occurring in Sichuan, Xizang and Yunnan. Its single species, *S. remalana*, shares a characteristic of numerous alpine plants in being densely hairy – presumably a protection against dessication. In other plant families the same end can be achieved through succulence or by an outer waxy layer.

METACHLAMYDEAE

Bignoniaceae

Typically tropical and warm–temperate, the family is well known in horticulture through a wide range of ornamentals including *Bignonia, Campsis, Catalpa, Eccremocarpus, Jacaranda, Pandorea* and *Tecomaria* and in temperate regions would be treated as conservatory specimens. *Campsis* and *Catalpa* are hardy, *Pandorea* and *Eccremocopsis* are half hardy. The family has no representatives, for example, in the British native or naturalised flora. It is therefore a surprise to find *Incarvillea*, a montane genus, well represented in Xizang, Yunnan, Sichuan, Gansu and Quinghai. *I. younghusbandii* is found at altitudes up to 5400 m in Xizang and Quinghai. *Incarvillea* provides one instance of a hardy member of this family adapted to Western horticulture.

Gesneriaceae

This family shares with the previous one a primarily tropical distribution but, as pointed out earlier, with alpine outliers more so in China than in Europe. High–mountain genera include three endemics, shown in Table 8.1. Of the 11 species of *Ancyclostomon*, 12 of *Isometrum* and 7 of *Tremacron*, the large majority are alpine.

Primulaceae

Among alpine plant enthusiasts, this family enjoys high esteem, notably, of course, for its principal genus, *Primula*, and which is considered separately in Chapter 11. Others of its alpine genera are worthy of comment, not least *Cyclamen* which is, so to say, an absentee being confined largely to the area from Iran through to Western Europe.

Pomatosace, found in mostly northwest China, grows at between 2890 and 4500 m. It occurs as one species *P. filicula*, not well known, but apparently existing in annual or perennial biotypes.

The genus *Androsace* occurs in the New and Old World but with its chief distribution in the latter. Its most widespread species is *A. chamaejasme* but among the more soughtafter Chinese representatives are *A. alchemilloides, delavayi, tapete* and *wardii*. A problem with growing *Androsace* in an ornamental rock garden is its tendency toward straggly growth by means of stolons that spring out and root at some distance from the mother plant. In nature, this serves as an effective means of increasing occupancy of a habitat. Collectors thus prefer the more compact cushion–forming variants.

Scrophulariaceae

Through such genera as *Antirrhinum, Calceolaria, Digitalis, Mimulus* and *Verbascum* botanists in temperate countries become aware of what is a cosmopolitan family and, as is evident, it is well represented in China. Among its seven endemic genera, three *Pterygiella, Scrofella* and *Xizangia* occur in alpine locations. The first of these has four species and the other two are monospecific.

A portion of this family, variously designated, but including such genera as *Bartsia, Euphrasia, Melampyrum, Pedicularis* and *Rhinanthus*, is semiparasitic upon other flowering plants. *Pedicularis* in China is a significant item in the alpine flora and several species, *P. corymbifera, longiflora* and *megolochila*, are of striking appearance.

Solanaceae

In no sense conspicuously alpine, this large cosmopolitan family is well represented in China. Of the two endemic genera here, one, *Atropanthe*, occurs at altitudes of up to 3000 m and the other, *Przewalskia*, is confined to altitudes between 3200 and 5200 m – a conspicuously high alpine. Both are used medicinally. The other montane genus, *Mandragora*, has gathered various legends round it. Arber (1990) both records the superstitions about this plant (in its European representatives) and, also, recalls early botanists who, commendably, put science before superstition. Her book, too, is worth reading for the light it sheds on early European botany.

Monocotyledons

Araceae

Although typical of wet or damp locations, the family is recognisably global. Apart from *Pistia* (water lettuce), a floating plant of the tropics, the family is normally instantly recognisable by the common characteristics of *Anthurium, Dieffenbachia, Dracunculus Monstera* and *Philodendron*. The inflorescence arising on a scape produces a spadix of tightly compressed small flowers, hardly recognisable as such with the naked eye. Around the spadix is a spathe, often conspicuously coloured, and in some species drawn into a long tail as in, for instance, *Arisarum proboscideum*.

Arisaema is noteworthy because even though it is characteristically a tropical genus, none the less, its species can be found in alpine areas of China at altitudes of up to 4000 m. There is thus a contrast with Bignoniaceae and Gesneriaceae, for example, where primarily tropical families have cold–tolerant genera. In *Arisaema* it is the genus itself which produces a range of about 150 species differing among themselves in cold tolerance and spanning such contrasted habitats. *A. griffithii* is Tibetan and *A. utile* grows in both Tibet and upland Yunnan. Lang et al. (1997) compare the spathe of the latter in appearance to an erecting cobra.

Liliaceae

As mentioned earlier various genera included in this family have been gathered, by some authorities into separate, newer families. *Asparagus* and *Smilax* were mentioned in Chapter 6. One such alpine plant in China is *Streptopus* – at first sight similar to either of the above genera but other than to Liliaceae, variously attributed to Convallariaceae or Uvulariaceae.

The lily family is cosmopolitan and, of course, conspicuous in horticulture with *Aloe, Colchicum, Hemerocallis, Hosta, Tulipa* and many more genera.

As on European mountains, Liliaceae is well represented in China. Apart from various *Lilium* species such as *L. duchartrei* there is the sought–after *Notholirion*, a small genus of six species occurring from Iran into west China.

Orchidaceae

Chinese orchids were discussed in the previous chapter. Table 8.1 indicates some, though not all, of the alpine genera. There are only two high-altitude genera endemic to China, *Porolabium* (Shanxi) and *Smithorchis* (Yunnan), each having one species.

Attention here is drawn to *Cypripedium* (the lady's slipper orchid group), differing from other alpine representations in having two anthers rather than one per flower, powdery pollen and in not forming pollinia. Although, at present, relatively widely distributed in the uplands of several provinces, notably Sichuan, Xizang and Yunnan, there is little doubt that, with an increase in tourism, these extraordinary plants will become even more of a target for acquisitive plant hunters. The matter is raised again in Chapter 15. *C. tibeticum* additionally, for example, has medicinal claims made for it.

Poaceae

As might be expected, among the grasses, it is the subfamily Pooideae which is best represented at high altitudes and where, in China, one finds *Poa, Puccinellia, Stipa* and *Trikeraia,* for example. Given that bamboos are normally found from warm–temperate through to tropical regions, it is noteworthy that species of *Sinarundinaria* occur in Xizang in the understorey of pine forest at about 3000 m. For details of upland vegetation see Chang (1981, 1983) and Zhang (1998).

Zingiberaceae

Normally thought of as Indo-Malaysian tropical, Zingiberaceae provides in addition to *Zingiber*, the ginger of commerce, *Alpinia* (coral ginger) and *Hedychium* (ginger lily), both striking ornamentals requiring a frost-free environment although it is sometimes possible to maintain *H. densiflorum* outside in southern England.

In contrast to the family generally, one genus endemic to China *Pyrgophyllum* can be found up to 2800 m in Sichuan and Yunnan and another, a non–endemic in China, *Roscoea* at still higher altitudes in Xizang, for example. Two Chinese species, *R. cautleyoides* and *R. humeana* are grown by perhaps the more ambitious horticulturists in Europe.

Endemics and Alpines

Although endemics are found at high altitudes, they are most frequent at intermediate altitudes and for a perspective on this the reader is referred to the earlier discussion of endemism in chapter 3. The majority of genera growing in alpine parts of China have representation elsewhere, not only in such habitats. To take three examples, *Saxifraga* occurs across the world's north temperate regions and via the Andes to Tierra del Fuego in the New World. *Sisyrinchium* is essentially, in both Northern and Southern hemispheres, an American genus. *Tremacron* is a Chinese endemic with species across a range of habitats although restricted to Sichuan and Yunnan. For a global perspective on the alpine genera mentioned in Table 8.1 see Beckett and Grey-Wilson (1993, 1994)

Three genera with appreciable alpine representation remain to be considered and for this and other reasons are given a chapter each. These are *Anemone, Primula* and *Rhododendron* and reserved for Part III of the book.

A Digression on *Gentiana*

Among the dozens of genera included in this chapter, it is appropriate to select one in particular for more extended treatment – namely *Gentiana*. This is not because of its generally striking appearance but, rather, its spectacular speciation within a Chinese context. Although the genus occurs throughout the temperate regions both north and south, especially in upland areas, it is its concentration north of the equator and especially in China that must command the attention of the plant geographer and taxonomist. One significant drawback is the complexity of its cytology, since many species have numerous small chromosomes per cell, a situation made no simpler by the presence of presumed supernumerary chromosomes, (Halda, 1996). In China as elsewhere, new species continue to be described and among those that have been known for some time the process of re-evaluation continues. The following therefore is a summary of the situation rather than a detailed critical treatment.

Worldwide there are perhaps 450 species subdivided into groups of 15 sections, some of these being further divided into series. Not all of either sections or series are represented in China. Table 8.2 summarises the situation for 1988 but see, also, the earlier paper (Ho 1985).

As will be apparent in this book, where possible we present evidence comparing traditional taxonomic arrangements with newer evidence from molecular biology. Using nuclear ribosomal DNA, although on the basis of a relatively small sample of gentian species, (12 sections and 24 species), Yuan et al. (1996) offered support for current more traditional arrangements except in regard to section Stenogyne which they suggest, on their evidence, might be removed from the genus. Chondriophyllae, Cruciata and Pneumonanthe they see as closely related sections.

Table 8. 2. Sections, series and numbers of *Gentiana* species occurring in China.

(Ho 1988)

[a] sections	series	Species nos. for China
Otophora	Otophorae	5
	Decoratae	3
Cruciata		16
Monopodiae	Confertifoliae	4
	Verticillatae	7
	Ornatae	14
	Apteroideae	7
Pneumonanthae		4
Frigidae		15
Isomeria	Sikkiimenses	4
	Depressae	5
	Stragulatae	2
	Uniflorae	6

Phyllocalyx		2
Microsperma	Suborbisepalae	5
	Tetramerae	2
	Annuae	2
Stenogyne		11
Dolichocarpa		12
Chondriophyllae	Fimbriatae	15
	Rubicundae	9
	Linearifoliae	18
	Orbiculatae	11
	Humiles	22
	Pubigerae	4
	Capitatae	7
	Fastigiatae	28
	Napuliferae	4
	Piasezkianae	3
		247

[a] The sections in these tables correspond approximately to what are called 'sub-genera' by Halda (1996)

A minor revision of Ho (1988) was that of Ho and Liu (1990)

Table 8. 3. The major geographic representation of *Gentiana* sections within 'Sino-Himalaya' with species numbers.

(After Ho and Liu 1990)

Sections (on a world basis)	Mts of S.W. China	Central and Eastern Himalayas
Otophora	10	5
Cruciata	10	3
Monopodiae	25	8
Frigidae	15	3
Gentiana	-	-
Pneumonanthe	-	-
Phyllocalyx	1	1
Calathianae	-	-
Ciminalis	-	-
Isomeria	15	9
Microsperma	9	2
Stenogyne	9	-
Dolichocarpa	6	5
Chondriophyllae	88	23
Fimbriata	2	2
	190	61

Table 8.3 offers a different perspective. Two areas in China of gentian concentration are compared. What tentative conclusions might we draw? Pneu-

monanthe *does* occur in China but not in these two areas, and so presumably has not been able to exploit and retain a place with the ecological opportunities on offer.   To this section those of Monopodiae and even more so, Chondriophyllae provide great contrast. Phyllocalyx is, also, instructive – holding on but hardly able to flourish.

What is it among the Chondriophyllae group of gentians that has prompted speciation not in one, but in both areas under consideration? What causes the Monopodiae to do well in one area but less so in the other in terms of speciation? These questions, applied here within one genus, can be extended to different genera and even families. The situation highlighted here for gentians is a microcosm of the wider situation among plants generally. We need to gather facts, but that is only the beginning of scientific progress. Seeing the same facts in different ways is a means of prompting new questions for which we have to frame new answers – a theme, which runs through chapters 10 to 13.

9 Medicinal Plants and the Meeting of Two Traditions

"...For since professional practice chiefly consisted in giving a great many drugs, the public inferred that it might be better off with more drugs still, if only they could be got cheaply".

George Eliot, *Middlemarch* (1871)

"Now on modern ideas one would expect that all the most powerful drugs, whether botanical, animal or mineral, to have been grouped together in the princely class, but this was not so at all; the mentality of the ancient Chinese naturalists was more sophisticated than that, health and hygienic - minded, less pharmacodynamic, so to say. For the princely drugs were defined as those which were good for general health, containing no dangerous principles and capable of being taken constantly without untoward effects. The adjutant drugs, on the other hand, were available for therapy in acute infections, contained dangerous active principles, had to be prescribed in small doses, and should not long be continued. The ministerial drugs occupied an intermediate position."

Needham et al (1986)

It is evident that Western and Oriental medical traditions differ, and it is largely beyond the scope of this chapter to explore in detail their philosophical bases. For that purpose the reader is referred to Needham et al (1986) quoted above or, more recently, Mills (1996). The objective of the present chapter is to consider the plants themselves, the active principles they contain and the extent, in selected examples, to which Chinese plants have added to the resources of Western medicine. Even so it will be necessary to indicate some points of contrast between Western and Oriental traditions. Beyond this, since the herbs used in Chinese medicine are gathered rather than farmed for the most part, demand tends to exceed supply and the consequences for conservation will eventually be considerable.

A Cautionary Tale

About folk medicine there can be a range of opinion. At one extreme it can be regarded as a panacea and at the other it is seen as deserving only unqualified scepticism. It is appropriate, therefore, to recount an incident from clinical experience in London. The essentials are as follows. For a fuller account see Phillipson (1994).

At Great Ormond Street children's hospital, patients were being treated conventionally for atopic eczema. Some patients improved and this was traced to the work of a traditional Chinese medical practitioner in central London. The Great

Ormond Street specialists made contact with this practitioner who had prescribed a preparation made from no fewer than ten different herbs. It was agreed among those practitioners involved representing the Western and Chinese traditions that, following systematic experiment, the results should be assessed following accepted criteria of Western dermatology. Inherent in this approach was the assumption that an active principle might be located in one of the herbs, the others being shown to be superfluous.

Table 9. 1. TCM prescription for severe atopic eczema in adults

Botanical identity	Latinised pharmaceutical	Chinese equivalent	TCM Category
Ledebouriella divaricata (Umbelliferae)	Radix ledebouriellae	Fang Feng	'Emperor'
Schizonepeta tenuifolia (Labiatae)	Herba schizonepetae	Jing jie	'Prime minister'
Rehmannia glutinosa (Scrophulariaceae)	Radix rehmanniae	Dihuang	'Minister'
Paeonia lactiflora (Ranunculaceae)	Radix paeoniae rubra	Chiahao	'Minister'
Lophatherum gracile (Poaceae)	Herbalophatheri	Danzhuye	'Minister'
Tribulus terrestris (Zygophyllaceae)	Fructus tribuli	Jili	'Senior officers'
Dictamnus dasycarpus (Rutaceae)	Cortex dictamni	Baixianpi	'Senior officers'
Clematis armandii (Ranunculaceae)	Caulis clematidis	Chuan-mutong	'Senior officers'
Glycyrrhiza uralensis (Leguminosae)	Radix glycyrrhizae	Gancao	'Junior officers'
Potentilla chinensis (Rosaceae)	Herba potentillae	Wielingcai	'Junior officers'

Eventually, and perhaps unexpectedly, it was shown that there was no single active herb or active principle and that the beneficial effects were due to the mixture of herbs. It is important to recognise that the results were derived from double blind[1] trials – one set for children and one for adults and that significant improvement was seen only in patients not taking the placebo. Table 9.1 presents the plant material so as to link Western and Chinese approaches.

[1] A double-blind trial is one where (a) the drug and the placebo are indistinguishable, i.e. similar capsules, for example, and (b) the person administering the drug or placebo does so not knowing which the controlling investigator has supplied.

Phillipson (1994) concludes:

"There are at least twelve different classes of chemical compounds present in the herbs and to date we have been able to isolate some 43 compounds from four of the herbs. A range of pharmacological activities has been reported for individual herbs in the prescription and these include anti inflammatory, anti-allergic, anti-bacterial, immune modulation, neuromuscular blocking and peripheral vasodilation. In order to ensure safety and efficacy of the herbs, it is essential that quality assurance procedures be used to standardise the herbs. Chemical tests and assay procedures are required in order to supplement the standards which are described in the Chinese Pharmacopoeia"

From the foregoing, together with this concluding comment from Phillipson it is evident that any judgement on traditional or folk medicine with its prominent herbal approach should at least pause to consider the evidence. In a later article, Phillipson (1995) places this work in a wider context of advances in plant chemistry in relation to folk medicine.

Utilising the Flora

In Chapter 1, reference was made to the Chinese pharmacopoeia tradition. In recent times, the aim has been to gather this and later material into the national pharmacopoeia, for example. Table 9.2 provides no more than a small sample of the 6000 or so species reckoned to be used. The following points are noteworthy.

1. The families are predominantly dicotyledons.
2. There is no obvious concentration of related families.
3. Unrelated families can be used to treat similar conditions.
4. Species found here for their medicinal properties include some listed as famine foods in the *Chiu Huang Pen Tshao* (see Appendix 1). (This is not unknown elsewhere and for example cloves used in the West for flavouring can also alleviate toothache).
5. The fifth column of the table indicates some at least of the active constituents and provides a link to the analytical tradition of the West.

Chinese Medicine and the West

The customary approach to any scientific problem in the West is to assume not only that effects have causes but also that these latter can be elucidated by appropriately designed experiments. Published conclusions are subject to rigorous challenge and medical science expects the same kind of analytic scrutiny as do the physical and biological sciences[1]. The prevailing approach has itself come under increasing scrutiny and alongside of it had developed a body of practice

---

[1] The Royal Society, the world's oldest such organisation, appropriately has for its motto *Nullis in verba* – 'take nobody's word for it'

Table 9. 2. A sample of Chinese medicinal plants (After Tang and Eisenbrand 1992 augmented by Williams pers. comm.).

Fl - flowers, Fr - fruit, L - leaf, R - root, Rh - rhizome, S - stem, T - tuber.
[a] Additionally listed as famine foods. See Appendix I

Family	Genus	Species	Organs	Active Chemicals	Uses
Araliaceae	*Panax*	*ginseng*	R	Ginsenosides (saponins)	
Aristolochiaceae	*Aristolochia*	*debilis*	Fr, L, R, S	Aristolochic acids	Antirheumatic, diuretic. oedema.
Asclepiadaceae	[a]*Cynanchum*	*stratum*	R	Glycosides	Antipyretic, diuretic
Caprifoliaceae	[a]*Lonicera*	*japonica*	Fl, S	Linalool	Emetic
Compositae	*Artemisia*	*annua*		Sesquiterpenes	Antimalaria
	[a]*Carpesium*	*abrotanoides*	Fr	Lactones etc.	Antifungal, antibacterial
	[a]*Carthamus*	*tinctorius*	Fl	Carthamin	Menorrhalgia
	[a]*Eupatorium*	*formosanum*		Sesquiterpenes	Anticancer
	[a]*Inula*	*britannica*	Fl	Gaillardin	Diuretic, mycolytic
Cucurbitaceae	[a]*Luffa*	*cylindrica*	Fr, L, S	Saponins	Stimulates interferon
	[a]*Trichosanthes*	*kurilowii*	R	Trichosanthin	Abortifacient, treatment of HIV, antianginal
Ephedraceae	*Ephedra*	*intermedia, equisetina sinica*		Ephedrine	Asthma, hayfever
Ericaceae	*Rhododendron*	*molle*			
Fumariaceae	[a]*Corydalis*	*incisa*	T	Alkaloides	Analgesic
Gentianaceae	[a]*Gentiana*	*scabra*	R	Gentianine	Anxiolytic, chloretic, urinary

Family	Genus	Species	Part	Active constituents	Uses / Actions
				Gentiopictin, Gentisin	Antiseptic.
Gesneriaceae	*Rehmannia*	*glutinosa*	R	Iridoid glycosides	Antipyretic haemostatic
Ginkgoaceae	*Ginkgo*	*biloba*	L	Ginkgolides, Ginkgflavones	Improving circulation especially in the brain, anti-asthmatic
Hydrangaceae	*Dichroa*	*febrifuga*	L		Antimalarial
Leguminosae	[a]*Albizia*	*julibrissin*	Fl, S	Saponins	Sedative
	Astragalus	*complanatus*			Sedative, hypnotic
		membranaceus		Betaine astragalia, sterols	-
Liliaceae	[a]*Sophora*	*japonica*	Fl, Fr	rutin	Antihaemorrhaigic
	Veratrum	*nigrum*		alkaloides	Emetic, insecticides
Moraceae	[a]*Morus*	*alba*	Fl, R, S		Antipyretic
Nelumbonaceae	[a]*Nelumbo*	*nucifera*	L	isoquinoline alkaloids	Anti tumour
Nyssaceae	*Camptotheca*	*acuminata*	Fr	camptothecine	Antipyretic, antiinflam.
Oleaceae	[a]*Forsythia*	*suspensa*			Antipyretic, vasostimulant, antitoxaemic
Paeoniaceae	*Paeonia*	*suffruticosa*	Bark	paeonolide paeonoside	Oedema, abscess
Periplocaceae	*Periploca*	*sapium*			Coughs
Phytolaccaceae	[a]*Phytolacca*	*acinosa*	R	saponins	Aiuretic
Polygalaceae	[a]*Polygala*	*tenuifolia*	R	triterpenes, saponins	
Polygonaceae	[a]*Polygonum*	*aviculare*	Fl, S		
	[a]	*cuspidatum*	R, Rh	anthraquinones	

Family	Genus	species	Part	Constituents	Uses
	a	*multiflorum*	T	oxygenated flavones	Tonic
	a	*orientale*	Fr	"	
	Rheum	*palmatum*	R	Anthroquinones including rhein and emodin	-
Ranunculaceae	[a]*Clematis*	*chinensis*	R	saponins	Antirheumatic, analgesic
Rubiaceae	[a]*Rubia*	*cordifolia*	R, Rh	anthraquinones	Antidiarrhoea
Simaroubaceae	*Ailanthus*	*altissima*		quassinoids	
Solanaceae	[a]*Lycium*	*chinense*	Fr	sesquiterpenes	Lower blood pressure
Theaceae	*Camellia*	*sinensis*	L	xanthines	Antiflammatory, choloretic, stimulant
Umbelliferae	[a]*Bupleurum*	*falcatum*	R	saponins	
	Peucedanum	*decurrens*	R	coumarins?	Expectorant, haemolytic
Verbenaceae	*Vitex*	*negundo*	Fr, L	flavonoids?	Antiseptic, disinfectant

For a more extensive list, see Mills (1996)

subsumed under the term alternative medicine. Acupuncture, aromatherapy and homeopathy, for example, are nowadays familiar terms and the establishment of immigrant Chinese communities has meant that Westerners can, if they wish, have access to oriental medicine. Quality control and the judicious presentation of realistic claims are matters that prompt official interest at the interface of the two cultures.

The Western tradition, partly through medical missionaries and partly through Chinese studying abroad, has been exported to China and, not surprisingly, there is the question of how far Western and traditional approaches there can be harmonised to provide an efficient and widely available medical service. For the moment, and probably indefinitely, most Chinese rely on their indigenous medical tradition.

Experiment or Experience

In a system of medical practice developed over millenia, the Chinese tendency has been to conserve and pass on a store of knowledge rather than generate a continuously innovative approach. Experience, more than experiment, has been the characteristic. The quotation from Needham et al at the opening of this chapter defines an expressly non-Western approach. While there are Westerners prepared to accept Chinese medicine on its own terms, this is not easy for the conventional experimentalist. For this latter, talk of *yin* and *yang* is at best seen as a distraction and, at worst, an irritating obfuscation. Rather than arbitrate in such matters, attention in this chapter concentrates on the use made of the flora.

There is a further point of contrast of which one needs to be aware. Normally any plant material used herbally will contain more than one active principle. Moreover mixtures of herbs rather than one prescribed species are commonplace. Given additionally the kind of variation encountered, whether of genetic or environmental origin, the problems of ascertaining statistically acceptable results are formidable. If we then add to this the tradition in the West that officially approved drugs are single molecules which have passed a defined series of procedures, it is obvious that the benefits of traditional Chinese medicine are not easily assimilated into Western practice. This is not to say attempts are not being made but rather to emphasise that the process is necessarily slow and expensive. Interest, however, continues to grow and the most helpfully illustrative way is consider a series of examples at different stages in the process of Western scrutiny.

Not surprisingly, the most recent work is unlikely to be in the public domain, due to the commercial considerations of various drug companies, and the following account depends only on published literature.

Medicines or Foods?

From a Western point of view, it has become practice to distinguish between medicine and food (or dietary) supplements. The latter category covers a wide range of materials, which although having a place in traditional or folk medicine, have no real claim to be foods. Since many of these are physiologically active toward humans, they are regarded more appropriately pharmacologically than nutritionally. Israelson (1997) has argued within the United States for a new category traditional medicine to be split off from food supplements. The United Kingdom viewpoint (Anon. 1996) is as follows.

"Food supplements cannot legally, to be capable of treating or curing human disease. They should be of 'appropriate quality' but this is not defined nor is there compulsory testing for safety. Trading standards officers can and do proscribe companies, which break these rules, but these are not drawn up specifically for dietary supplements and can be difficult to enforce. Although the Medicines Control Agency can take action against producers it does only in response to a complaint".

See also Anon. (1998). Recognisably, advocates or devotees of one tradition can be uncomfortable with a regulatory system designed for the other. It is not merely a matter of dialogue among practitioners. The sick, wanting to be made well and ready to spend much in order to achieve this, fuel interest and indeed anxiety. Where there is a demand, there will be suppliers, and should it be the entrepreneur or the analyst who decides the outcome?

With these considerations in mind how, and to what extent, can traditional Chinese medicine be incorporated into Western clinical practice?

The Ladder of Adoption

It is possible to identify a series of essential stages in the process from recognising a potentially useful source to its adoption, clinically, in Western medicine. It must be analysed chemically to identify active constituents. These are each then made the subject of a bioassay involving, eventually, testing on animals. From there a preparation in some form is given to healthy humans, and subsequently, in a double–blind trial to patients suffering from the target illness. At each stage various medical and ethical safeguards are in place. There is, so to speak, a "ladder" set up, each step being an essential prerequisite to the one above. It is possible to identify in this way two things. The first is whether some traditional remedy is near to the top of the ladder pending its adoption, or whether it is regarded merely as a food supplement, for which no claims are made even though by implication the purchaser is led to regard the material as health-improving or health-sustaining. For a more elaborate treatment of such a ladder see Beal and Rienhard (1980)

Fig. 9. 1. A sample ladder of adoption

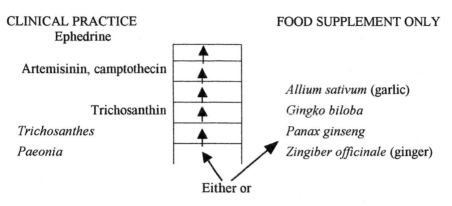

CLINICAL PRACTICE   FOOD SUPPLEMENT ONLY
    Ephedrine

Artemisinin, camptothecin

                                  Allium sativum (garlic)

        Trichosanthin   *Gingko biloba*

Trichosanthes   *Panax ginseng*

Paeonia   *Zingiber officinale* (ginger)

Either or

Ephedrine

This is an alkaloid isolated many years ago from *Ephedra vulgaris*. It is a sympathomimetic and decongestant and increases blood pressure. It has been used in the treatment of asthma. As regards Western medicine, its science has moved on and ephedrine has largely passed out of use, being replaced by more satisfactory alternatives.

Artemisinin

The following account is based on Klayman (1993)

For about 20 centuries the Chinese have recognised that a tea made from *Artemisia annua* (Compositae) could alleviate fever. A prescription is available from A.D. 340 in the *Handbook of Prescriptions for Emergency Treatments*.

In modern times, from 1971, it was known that a low-temperature extraction with diethyl ether gave a substance having antimalarial activity against infected mice and monkeys. The active principle was shown to be artemisinin, a sesquitepene lactone incorporating an endoperoxide moiety.

Of the 300 or so *Artemisia* species, artemisinin is virtually confined to *A. annua* occurring at about 0.94% w/w in China with lower values in the United States. It is significant that, in this respect, traditional Chinese medicine had identified precisely the one species shown to be of any real use in this situation.

In 1971 artemisin was shown in China to be effective against *Plasmodium vivax* and *P. falciparum*, including some variants resistant to chloroquinone.

On present evidence, there is every reason to believe that artemisinin and some of its derivatives comprise an important weapon in the control of malaria. See also Beakman et al (1998), where such chemicals are considered in relation to cancer.

Camptothecine

Camptotheca acuminata (Nyssaceae) is known in Chinese medicine as an anti-tumour plant. From it camptothecine has been isolated and this is now an established anticancer agent. For a detailed review see Blasko and Condell (1988) and more recently Wall and Wani (1993).

Trichosanthin

From *Trichosanthes kurilowii* (Cucurbitaceae) a substance, trichosanthin, has been obtained. Traditionally the plant has been used as an abortifacient and more recently against HIV, and although it is uncertain with what degree of effectiveness, it does appear to have some activity in this direction.

The material trichosanthin is a ribosome-inactivating protein and its abortifacient activity is due to its high toxicity against trophoblasts. It has also been shown under in vitro conditions to damage choriocarcinoma cells.

Paeoniflorin

It will be recalled from Table 9.1 that one ingredient in treating atopic eczema was *Paeonia lactiflora* (Paeoniaceae). It has been shown that this peony will yield paeoniflorin and, too, a complex glucopyranose. The precise contribution these might make in the cure of that disease is not yet clear. Mills (1996) lists this species as "tonic, analgesic, spasmolytic"

The other items shown in Fig. 9.1, although shown as food supplements, have prompted clinical interest. Of those perhaps the most beckoning is *Ginkgo*, where claims are made for its ability to arrest the decline of mental function and which will be considered shortly. For those in both East and West committed to the virtues of *Panax ginseng*, this species, also, is examined later.

Diseased Plants

Healthy plant tissue as a part of normal metabolism generates a retinue of chemicals - some pharmacologically active and, given the number of species available, a practically inexhaustible supply. Even so, to these can be added chemicals generated only in diseased tissue. Perhaps the best known is the generation, in rye, of alkaloids such as ergotamine when infected with the fungus *Claviceps purpurea* (Clavicipitaceae) and used to stimulate contraction of the uterus.

More recently an instance is provided by *Rehmannia glutinosa* (Scrophulariaceae) infected by *Aspergillus, Fusarium, Candida* or *Saccharomyces*. Infected, but not healthy tissue, generated two compounds (phenolic glycosides) which on isolation were found to be active against *Pseudomonas capacia* and *P. multiphilia* (Shoyama *et al* 1987).

Ginkgo biloba

It is appropriate here to add a brief section on this curious gymnosperm, the maidenhair tree, familiar in the West as an ornamental.

Medicinally, current interest centres on the real or supposed effects of ginkgo leaf extracts upon human mental capacity. Given the increasing life expectancy in Western societies, a significant proportion will succumb to severe mental disfunction in, for example, Alzheimer's disease. Since such a condition can create both severe stress among carers and loss of dignity, progressively, for the sufferer, any ameliorative or preventive treatment can attract clinical exploration. In the West, leaf extracts of ginkgo have been used in such studies. There are, even if such extracts are shown to have no harmful side effects, obvious risks of sensationalist reporting and the rousing of unreal expectations.

For about two decades work in the West has taken significant interest in ginkgo and various claims have been made. It was necessary to challenge the methodology. Kleijuen and Knipschild (1992) reviewed critically some 40 studies on the effects of ginkgo extract on mental impairment in the elderly. In the view of these authors only eight of these studies showed an acceptable level of experimental rigour. Although placebos were used, none incorporated double blinding. Their conclusion, despite this shortcoming, was that "there seems to be some biological plausibility for the potential beneficial effects of ginkgo". They go on to affirm that, given the absence of clear side effects, the authors themselves might try the treatment using leaf extract. They do, however, call for larger numbers of patients, effect measurement and improved experimental design.

Work has continued on the use of ginkgo in this area in both alternative and mainstream medical research. See for example Curtis-Prior et al (1998) and Dore et al (1999).

In a rather different direction Rigney et al (1999) explored the effect of ginkgo leaf extract on volunteers between 30 and 59 years of age who were asymptomatic, that is showing no signs of mental impairment. The trial involved both placebos and double–blinding. Evidence of shortening of reaction times and of a clearer rate of improvement in older volunteers emerged. Some 36 volunteers participated in the study, presumably a non-random sample in that they might be expected to have a positive interest in their own and, perhaps, other people's mental capacities.

Ginkgo extract is not the only chemical in which such interest exists but it is an intriguing ingredient in probing both declining and sustained mental capacity, an area of vast concern in modern medicine. Not surprisingly, the understanding of *Ginkgo* metabolism and development is now being advanced. See, for example, Balz et al (1999).

Panax spp. (Ginseng)

Ginseng is perhaps the best known by Westerners of Chinese medicinal plants, and in China itself the plant is, curiously, esteemed as both a tonic and a sedative. The physiological effects of ginseng are the subject of considerable debate and experiment and it is beyond the scope of this book to do much apart from raising certain issues. Materials supposedly or genuinely derived from ginseng are available commercially and a report from Sweden deserves consideration.

Cui et al (1994) examined a range of ginseng products, of which some were devoid of ginsenosides, the active chemicals characteristic of *Panax ginseng*. One of the preparations actually contained ephedrine and led to one athlete who had taken it testing positive for doping. Such a situation does underline the need for the maintenance of quality standards.

Three species of *Panax* are of principal interest *P. ginseng* (Asian), *P. quinquefolius* (American) and *P. japonicus* (Japanese) ginseng. From all three species it has become apparent that there is wide variation for each in the content and type of constituent ginsenosides and, also, depending on growing conditions, the influence of the environment is an important factor. This helps explain how a ginseng preparation depending on its origin might be, for example, either a tonic or a sedative to put the matter at its simplest. For a review of this complex topic see Attele et al (1999).

Concluding Comments

For recent discussions of the interaction between Western and Chinese medicines see the following among many that this subject has generated in recent years. Williamson et al (1996), Read (1993), Xiao (1994) and Zhou (1995).

The subject area indicated by this chapter is immense and growing rapidly. As an indication of this, the Royal Botanic Gardens, Kew, in the United Kingdom, has recently established an international authentication centre for traditional Chinese medicine.

It seems not unreasonable to point out that Chinese plant life is under pressure for several reasons, not least because so many species used medicinally are gathered from the wild rather than farmed – a situation which needs, surely, to be reversed and to which further reference is made in Chapter 15.

Part III

Some Major Genera

10 *Camellia*

Camellia, a genus native to Indo Malaysia, China and Japan, contains an uncertain number of species variously estimated at between 80 and 300. It was introduced to Europe in 1698 by James Cunningham and from there, according to Bennet et al (1995), had reached America through the efforts of John Stevens by about 1797. Since part of the genus is substantially hardy. its place in temperate gardens is assured, producing, as it does, spectacular flowers from February to May, for example, in southeast England. In warmer countries a selection of varieties will ensure year-round availability of flowers.

Camellia sinensis, the tea of commerce although grown best in relatively cool climates the temperature should not drop below about 12–13 °C. Tea has a substantial technical and popular literature of its own. In the interests of a wider appreciation of the genus it is treated only marginally here. In China, tea drinking was first recorded in the *Ehr Ya*. The earliest ornamental camellias imported were red, pink and white types. Yellow camellias arrived and were described as *C. flava* and *C. tonkinensis* in 1910 (Savidge 1994). They and other yellow species tend to be less hardy and attempts are underway to transfer this colour to more robust germplasm.

Theaceae, the family to which *Camellia* belongs contains other ornamental genera, notably, from China, *Anneslea, Gordonia* and *Schima* together with *Apterosperma, Eurya, Stewartia* and *Ternstroemia* from mostly elsewhere. Two other genera, *Hartia* and *Polyspora*, are sometimes considered sufficiently distinctive to merit segregation from *Stewartia* and *Gordonia*, respectively. These latter genera, together with *Camellia*, comprise the subfamily Theoldeae.

For camellias, as with other ornamentals, interest is driven largely by a commitment to produce new saleable varieties. The result is a concentration upon only parts of the genus. The present chapter, therefore, is concerned with the wider botanical context, which under the impact of newer techniques, plant taxonomy is beginning to change. For several reasons the genus *Camellia* provides, in these circumstances, a convenient initial example of how such changes are handled.

Camellia Classification: the Traditional Phase

The steady arrival of new specimens from the Far East creates two problems. The first is to ascertain whether the material is genuinely novel or a replicate

Fig. 10. 1. Sealy (1958)   (After Parks, 1992)

Section

I Archecaellia	II Sterocarpus	III Theopsis	IV Camelliopsis	V Piquetia	VI Thea	VII Corallina	VIII Calpandria	IX Pseudocamellia	X Heterogenea	XI Camellia	XII Paracamellia
C. chysantha		*C. cuspidata* *C. fraterna*	*C. salicifolia*		*C. sinensis*				*C. granthamiana*	*C. japonica* *C. reticulata*	*C. sasangua* *C. olifera*

Fig. 10. 2. Chang (1981)

Subgenus	I Protocamellia	II Camellia	III Thea	IV Metacamellia
Section	1. Archecamellia *C. granthamiana*	4 Oliefera *C. oleifera C. sanguinea*	11 Corallina	19 Theopsis *C. fraterna*
	2. Sterocarpus	5 Furfuracea	12 Brachyandra	20 Eriandria *C. cuspidata C. salicifolia*
	3. Piquetia	6 Paracamellia	13 Longipedicellata	
		7 Pseudocamellia	14 Chrysantha *C. chrysantha*	
		8 Tuberculata	15 Caloabdra	
		9 Luteoflora	16 Thea *C. sinensis*	
		10 Camelia *C. japonica C. reticulata*	17 Longissima	
			18 Glaberrima	

sufficiently close to that which we have already. The second is, as the numbers of genuinely new species accumulate, the need to put them into some kind of rational arrangement. We need a reliable system of pigeon holing. This implies a means of keying out which is accurate and, preferably, straightforward.

More ambitiously, one might then embark on arrangements which supposedly reflect "natural" relationships, that is, the underlying evolutionary trends.

Two traditional systems of *Camellia* classification, which have found wide usage, are those of Sealy (1958) and Chang (1981). To facilitate comparison between them they are set out in Figs. 10.1 and 10.2, incorporating data from Parks (1992). The following points emerge.

1. Below generic level, Sealy adopts a single and Chang a double layer
2. Names of sectional significance in Sealy can have subgeneric or (lower) sectional significance in Chang.
3. Park's instructive use of ten species allocated under both systems reveals little common ground.
4. Those *pairs* of species selected by Parks tend, by virtue of similarity, to appear together under both systems
5. Regardless of which might be the more natural system, each can serve a pigeon hole function choice of which is largely a personal preference

Given the coexistence of two such systems, there is eventually the need to arbitrate between them and if neither, in the long run, proves sufficiently useful, to devise a new system.

Cross Compatibility

An approach, sometimes appropriate, is to cross a group of species in all combinations to ascertain whether they cluster into subgroups on the basis of compatibility. Attempting this in *Camellia* might be feasible to discriminate between the systems of Sealy and Chang. In fact, an extended crossing programme reported by Parks (1992) produced no pattern which favoured either system.

To cross three species reciprocally requires (3 times 3) minus 3 operations, four species (4 times 4) minus 4 and 30 species (30 times 30) minus 30 or 870 operations minus discounting the essential replication required. Once more than about a dozen species are involved, the attraction of alternative methods, both more rapid and more informative, soon becomes apparent. Test crossing retains a place, but it is best confined to key species and used in conjunction with more modern approaches. See, for example, Parks et al (1995).

Newer techniques have been applied to *Camellia*, one of which will be considered here in some detail.

DNA: Some Background Considerations

DNA represents the genetic code defining any organism. Ultimately, the aim must be to compare DNA from different organisms on the basis that the greater the differences in their genetic codes, then the more is the evolutionary divergence among them. (Such an approach has the additional advantage that it enables us to set aside, at least temporarily, all other data, including morphology and which we have, necessarily, to view somewhat subjectively). Having recognised the significance of DNA from this viewpoint, one then requires a convenient source of DNA likely to be subject to certain ground rules and around which a suitable replicable technology can develop. Potentially, DNA could be extracted from the chloroplast, mitochondrion or nucleus, and it is the first of these that is now established as pre-eminent in helping resolve questions of flowering plant phylogeny. Attributes which make chloroplast DNA suitable include these five.

1. Total DNA content of a single chloroplast is of the order of 150 kilobases – a relatively small amount and thus conveniently handled.
2. Since the chloroplast mediates a process fundamental to green organisms changes to the organelle's DNA are likely to be on a relatively small scale and comparatively infrequent. The organelle is in this sense conservative
3. There is thus from one organism to another a general similarity providing a convenient context within which to view the relatively small changes which occur in either individual bases or short lengths of DNA
4. In eukaryotes (higher organisms) unlike prokaryotes (lower organisms – bacteria and blue-green algae) there occur in their genes non-coding or intron regions. Given the ability to sequence for long stretches of DNA changes in order and content here can complement data derived from point mutations involving single base pairs. (This is true of all three DNA–containing organelles but, as pointed out earlier, considerations on the basis of genome size apply to cp. (chloroplast) DNA.
5. Among flowering plants, the customary situation is that chloroplast inheritance, unlike that of the nucleus, is uniparental allowing one to follow the maternally donated organelle. (Different considerations apply among gymnosperms and were explored in Chapter 5).

For reviews see Palmer *et al* (1988) and Olmstead and Palmer (1994) for earlier and later accounts respectively.

So powerful is this technology now reckoned to be, that its use has spread rapidly. Increasingly, we are surprised if, for a genus of some importance, this kind of information is not yet available. Not unexpectedly, therefore, phylogenetic studies based upon cp. DNA are becoming available for *Camellia*.

An Application of cp. DNA to *Camellia* and its Relatives

Earlier, reference was made to the subfamily Theoideae, a grouping within the family Theaceae. To the Theoideae Prince and Parks (1997) applied cp. DNA techniques to test closeness of relationship. Two genes, *rbcL*, coding for part of the RuBisCo protein and *matK*, another protein–coding gene were used and applied to 30 taxa. What emerged was the relatively small divergences among species within a genus and only slightly larger divergences among genera. On this basis there is relatively little support for splitting off *Hartia* and *Polyspora*. It must, also, alert us to the prospect that we *might* have been too ready to grant specific status to various seemingly distinctive variants within *Camellia*.

Within-*Camellia* Variation

It is at this point appropriate to consider a study by Thakor (1997). Its approach is in three parts. It consists of an in-depth reconsideration of morphology with an emphasis on living plant material, an analysis of crossability data and an application of cp. DNA technology, all within the genus *Camellia*.

Preliminary reports so far available from crossability data do little to support either Sealy or Chang's system except for the Chrysantha section of Chang. Bringing together, eventually, all three strands of this investigation offers an absorbing prospect. For the present the arrangement of the genus *Camellia* remains provisional and unsatisfactory.

A Pragmatic View of *Camellia*

Thakor (1997) remarks:

"On the other hand, if taxa are only recently speciated, then their DNA may be largely invariable in areas of the genome unrelated to the speciation event. In such instances the apparent morphological differences could be due to mutations in one or a few genes active in some early developmental pathways and the remaining genome could be largely invariable among the taxa".

If camellias in their more conspicuous centres of diversity were confined to areas known to be geologically recent, one might suspect there only incipient speciation rather than something longer and more profoundly established. Again, if such variants retained a high degree of mutual crossability they could be regarded as relatively recent segregants. Such impressions would be reinforced if, in widely separated geologically stable areas, diversity was meagre and crossability uncommon. What light is there to be shed by systematic collection within China and to some extent in Japan?

While *Camellia* is spread through Indomalaysia, China and Japan, China contains the largest number of reported species and is, for this present book, the focus of interest. Parks (1995) remarks:

"The Japanese have thoroughly searched their hills for interesting variants among the wild populations of the four species that grow there. The diversity, however, among the more than 250 *Camellia* species native to southern China mostly remains to be explored. Each time I visit the Chinese countryside, I am amazed at the array of camellias that grow there".

What kind of information supports this view? It is now known, for example, that yellow camellias are more common than was once thought. To *C. nitidissima* can be added the yellow *C. impressinervis* and, moreover, the first species is more variable than seemed likely. To the fragrant species *C. lutchensis* can be added another, *C. henryana*. Not surprisingly, hardiness was detected in *C. reticulata* and *C. saluensis* at elevations beyond 3000 m, and *C. reticulata* is now known to have a range of chromosome numbers. Parks (1995) points out that in Yunnan "literally dozens of new *Camellia* species have been described in recent years!" We know Yunnan, in its present aspect to be relatively young geologically. Does chromosome number reveal anything of interest?

Chromosome number in *Camellia* was known to range from 30 to 90 with $x = 15$ (Darlington and Wylie 1955) but the range is wider. Table 10.1 summarises some more recent information.

One species, *C. forrestii* from Yunnan is present in $2x$, $4x$ and $6x$ variants. *C. pitardii*, also from the same province, has $2x$ and $6x$ variants. An octoploid is now known from Hunan.

Table 10. 1. Chromosome counts in *Camellia*

(After Gu et al 1989)

Yunnan		Guangxi		Guizhou		Hunan	Zheijiang
[a]19 spp.	15 $2x$ 2 $4x$ 7 $6x$	8 spp.	6 $2x$ 1 $4x$ 1 $6x$	5 spp.	3 $2x$ 2 $6x$	1 sp. $8x$	1 sp. $2x$

[a] some spp with more than one ploidy level

At its simplest, from classical genetics, one might assume that where a tetraploid species is fertile it is more likely to be allo- rather than autoploid. That is to say two, rather than one, species have provided its constituent nuclear genomes. For a hexaploid the likelihood of two species having contributed is higher and three is possible. If one then adds the proviso, as seems reasonable, "segmental" alloploidy probably complicates matters. The upshot is that the existence of fertile polyploids points to a convincing degree of previous diploid speciation. To provide supporting evidence requires that a tetraploid (say) is resynthesised from its putative diploid projenitors and shown to be cross compatible with the naturally occurring tetraploid. To date, no such resynthesis is known for any *Camellia* polyploid.

Camellia japonica

C. japonica is interesting as an outlier of the genus' distribution, especially as it has been closely studied. Essential aspects include the following. How *Camellia* reached Japan from China originally is unclear, but Miocene fossils of "*C. proto-japonica*" have been described by Yoshikawa and Yoshikawa (1996) from Honshu and Hokkaido. It does not follow, of course, that present-day *Camellia* in Japan is the lineal descendant of the Miocene fossils nor are repeated episodes of introduction precluded, some of which might have persisted through their offspring to the present day.

Given Japan's chequered geological history, Honshu and Hokkaido were themselves separated into smaller islands with, presumably, genetic consequences for fragmented plant populations. Later, under the impact of the Ice Ages, populations might have migrated in response to changing climatic patterns.

Wendel and Parks (1985), on the basis of isozyme studies, have shown for *C. japonica* high within-population diversity. As perceived by them, *C. japonica* has diversified into subsp. *japonica* widely distributed, subsp. *macrocarpa* of restricted southern distribution and subsp. *rusticana*, a more prostrate upland species also of restricted distribution. There is, however, debate as to whether subsp. *rusticana* should be given specific status (Yoshikawa and Yoshikawa 1997). Wendel and Parks (1995) refer to this and subsp. *japonica* as "clinally intergrading".

Native populations of *C. japonica* are currently reckoned to be challenged in three ways. Progressive urbanisation has brought with it diminution of habitat. Contrary to what might be expected, this has been thought likely to *increase* diversity due to cross–pollination from nearby new garden plantings of exotic selections. Thirdly, the tea plant *C. sinensis* has provided a further source of contaminating germplasm giving, subsequently, plants that resemble *C. japonica* but contain caffeine or resemble tea but with pink flowers (Yoshikawa and Yoshikawa 1997). These authors recommended that to trace authentic *C. japonica*, material should be taken from shrubs that were over 30 years old.

Given these various observations, the trading links, even if intermittent, for more than a millennium between China and Japan and the interest citizens of both countries have in *Camellia*, one hesitates to attribute genetic diversity in *C. japonica* in Japan to only recent human intervention.

Yellow Camellias Reconsidered

Fu and Jin (1992) list ten species of Camellia under threat – *C. chrysantha, euphlebia, tunghinensis, pingguoensis, pubipetela, crapnelliana, granthamiana, grijsii, reticulata* and *sinensis* var. *assamica*. Of these, the first five are yellow and confined to a relatively small area of Guangxi. Given the close similarities of these species and their location they provide useful material, perhaps, for a study of their speciation. Lu and Huang (1995) report hybrids between *C. nitidissima* (yellow) and *C. japonica* and between *C. tunghinensis* and *C. nitidissima*. In a separate study, Xia (1996) has reported the existence of 22 yellow species.

Concluding Comments

It is almost certain that, for a few years yet, new species of *Camellia* will continue to be described from China. In one sense this is to be encouraged if it enables us to become more aware of the genetic diversity present and how it might ultimately enhance the range of choice of new, even fragrant, yellow varieties. Instinctively, though, there are reservations; in particular it is necessary to resolve whether or not in the plethora of newly described species each genuinely deserves such a status. We could gain either way; if speciation is merely incipient, their use in breeding could lead more readily to new varieties. If speciation has led to genuinely distinctive taxa, then there is the challenge to understand how evolution could be so rapid given the likely time scale involved. With this in mind, we turn to different and contrasting genera primarily because each has its own pattern of biodiversity across a range of ecodiversity and, underlying these, its own distinctive genetic style of doing things. If, in the same young geological environment with its colossal array of microhabitats, one genus has rigidly defined species boundaries and little detectable natural (or even artificial) hybridisation and another has seemingly "fluid" genetic boundaries, how does this come about? Is it that one genus is "old" and another "young" or is there some other explanation? How is it that one genus produces few, and another many, species in the geologically young parts of China? What constitutes a different responsiveness in each to the myriad opportunities generated? How does one select the most informative genera to answer this kind of question? Above the generic level, do we find that families themselves are more or less responsive or do they, so to say, leave it to one or two enterprising genera among the rest? Do "primitive" families, arguably older, show less speciation than those more "advanced" and presumably younger? Alternatively, do we need a different frame of reference to understand and explain the extraordinary speciation in some, but not all, parts of China?

11 *Anemone* and *Primula*

Anemone

Depending on which taxonomists are consulted, the genus *Anemone* is reckoned to have, worldwide, from 120 to 150 species, distributed primarily in the north–temperate region, though with outliers on the upland parts of tropical mountains. *A. rivularis* occurs, for example, in Sri Lanka, India and west China. *A. sumakense* is found in Java, Sumatra and India.

Opinions differ as to the boundaries of the genus. Drawn narrowly it can exclude the taxa *Hepatica* and *Pulsatilla* or broadly, include them. For the present discussion, *Hepatica* and *Pulsatilla* are not merged with *Anemone*, but left until later for further comment.

The flower of *Anemone* is not quite what it seems. It lacks petals, their place being taken by coloured sepals. These, in turn, are subtended by bracts. The family to which it belongs, Ranunculaceae, is regarded as "primitive", a view tracing back to the American botanist Bessey (1915). For other reasons, *Primula*, considered later in the chapter, is regarded as "advanced". *Rhododendron* and *Rosa*, examined in Chapters 12 and 13, are thought, in this respect, to occupy intermediate positions. Among *Anemone* species most familiar in horticulture are *A. blanda* (Asia minor) and *A. coronaria* (southern Europe). *A. nemorosa* is a familiar denizen of deciduous woodlands and its variant Vestal is a double form. *A. huphensis* (syn. *japonica*), however, is from China.

It is convenient to take *Anemone* before *Primula* here since questions common to both are raised in simpler form with the first and smaller genus. In any genus, where species are numerous, it is convenient to group them on some basis, and that followed here equates with the *Flora Reipublica Popularis Sinicae*, (Wang 1980), shown in Table 11.1.

Of 53 species only about 13 have found their way in to western gardens in any quantity and of these only two are confined to China

Apart therefore from an few specialist collectors it is the case that most Chinese anemone species are unknown to most horticulturists. Of what then does this largely unknown group of *Anemones* consist?

Table 11.1. Sub division of the genus *Anemone* for species occurring in China

Section	No. of Species	Examples in horticulture
Anemonathea	12	*altaica*
Stolonifera	8	*baicalensis, flaccida*
Rivularidium	2	*rivularis*
Begoniifolia	4	
Anemone	8	*huphensis, rupicola, tomentosa, sylvestris, vitifolia*
Himalayicae	9	*obtusiloba, trullifolia*
Homalocarpus	9	*demissa, narcissiflora*
Anemonidium	1	
Total	53	

Distribution

One can recognise, for present purposes, three classes of *Anemone*, those occurring outside of China, those China shares with other countries and those peculiar (or endemic) to China. Of the 53 species indicated in Table 11.1 Table 11.2 considers only the 27 species endemic to China, grouping them with regard to selected provinces.

Table 11. 2. Distribution of endemic *Anemone* species in China

Province		Species numbers	
Sichuan	6	⎫	
Xizang (Tibet)	1	⎬ 12	
Yunnan	5	⎭	⎫
Sichuan + Yunnan	3		⎬ 23
Some or all of S, X and Y with other provinces	8		⎭
Other provinces excluding S, X and Y	4		
	27		

Table 11.2 shows that endemic *Anemone* species are significantly well represented in Sichuan and Yunnan. Were we to add the species China shares with other countries, we would find those more widely dispersed are appreciably represented in Sichuan, Xizang and Yunnan. These three provinces are thus "hospitable", it seems, capable of both generating and harbouring novelty.

The richness of Sichuan and Yunnan expressed here in *Anemone* is, as will by now be evident, a recurrent theme of Chinese botany. Yunnan, reaching from the tropical area near the Burmese border to the Himalayan snowline, is immensely diverse and is home to more than 14000 species, or just under half the Chinese flora. Sichuan, less geographically spread, has a rich concentration of species including many which have become familiar garden subjects. Even so, comparing *Anemone* with other genera, it becomes apparent that its richness even in Yunnan is not so rich as all that. The point will be explored subsequently in *Primula* and *Rhododendron*.

With regard to *Anemone*, it does not follow that unfamiliar species need be better or worse than those we already know in terms of interest to the horticulturalist. Some horticultural reappraisal of this genus as represented in China seems timely.

Pulsatilla

This genus, clearly so closely related to *Anemone* as to be included sometimes with it, provides a thought-provoking contrast. About 30 species occur, spread across appropriate habitats in Eurasia. Nine species occur in China, none seemingly endemic there and, relative to *Anemone*, more northerly in distribution. Not every genus or species can respond to new ecological opportunities. Was *Pulsatilla* a static, stabilised genus simply unable to respond genetically to, say, the rise of the Himalayas and the multiplicity of new habitats thereby on offer? The Chinese species are *ambigua, campanella, cernua, chinensis, dahurica, patens, sukaczevii, turezaninovii* and *millefolium*.

Hepatica

Ten species of this genus occur in temperate Eurasia. *Hepatica* is perhaps even less ecologically "enterprising" than *Pulsatilla*, although it is included for purposes of comparison when considering later, chloroplast DNA. The Chinese species are *H. henryi* and *nobilis*.

Generic or Subsidiary Status?

To botanists who have considered the matter, *Anemone*, *Hepatica* and *Pulsatilla* are recognisably distinct entities but not necessarily all deserving generic status. They could, for example, be set out in one genus, *Anemone*, with subgenera Anemone, Hepatica and Pulsatilla. On a morphological basis the differences are set out in Table 11.3.

That *Pulsatilla* and *Hepatica* are "small" genera compared to their close relative *Anemone* is a point of some interest and will now be explored.

Table 11.3. A morphological comparison among Anemone, Hepatica and Pulsatilla

	Anemone	Hepatica	Pulsatilla
Sepals	Petal-like	Petal-like	Petal-like
Cauline leaves	Well below the flower	Close to the flower Resembling a calyx	Well below the flower
Fruit	Achenes glabrous or pubescent with a beak	Achenes hairy and short-beaked	Achenes developing plumose tails

A View of Generic Size

Even when allowance is made for differences in taxonomic judgement about the number of species in a given genus, there are "large" and "small" genera. *Solanum*, worldwide, has perhaps 2000 species. *Rhoeo* has one and, within frost-free locations, despite its peculiar genetic system, is remarkably adaptable. Between these extremes, *Ficus* has about 900 and *Eranthis* seven species. These situations can be read in more than one way. Many species in one genus could mean, as with *Ficus*, given its pollination mechanism, exploitation of subsidiary niches. Elsewhere, as with *Solanum*, the genus is remarkable for its geographical spread, range of morphology and occupancy of diverse ecological habitats. A relatively small number of species in a genus could mean either that it was young and only beginning to radiate or that a few surviving relicts represented its demise. Alternatively, they might be relatively all-purpose species surviving indefinitely in a range of habitats.

Against this background one can recognise "intermediate" genera having, say, between 100 and 200 species. At this level we might expect a genus to occur in more than one phytogeographic region and to manifest, maybe, one or two contrasted "themes". At the very least, more than one level of ploidy would imply older species with lower chromosome numbers and younger polyploids derived from them. However, even this convenient generalisation need not hold if a group of diploids, recently presented with a newly derived environment, responded with a rash of speciation at the diploid level.

Behind all this is another consideration, which is whether different genera can respond equally to new situations, such as, for example, to the rise of the Himalayas. Was *Primula*, for example, more adept in responding than *Anemone* and, if so, how?

'Cytogenetic Repertoire'

In terms of what is visible of chromosomes by optical microscopy, a possible cytogenetic repertoire can be summarised for *Anemone* as in Fig. 11.1

Fig. 11.1. An interpretation of chromosome numbers in *Anemone* (After Darlington and Wylie 1955)

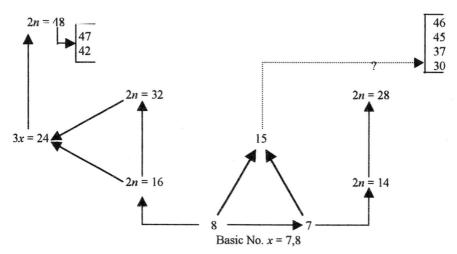

Basic No. $x = 7,8$

The simplest assumptions are that diminution of basic number from 8 to 7 is more likely than the reverse, fertile tetraploids are more probably alloploid rather than autoploid and that aneuploid chromosome numbers, not obviously multiples of the basic number, are more likely to be maintained by vegetative reproduction. How far these might apply to *Anemone* is then a matter for investigation.

Diploids on $x = 8$ and $x = 7$ occur in both the New and Old Worlds, as do tetraploids on each. An interesting experimental link between anemones from each was made by Gadjewski (1946, quoted by Stebbins, 1950). *A sylvestris* ($2n=16$) Old World was crossed with *A. multifida* ($2n = 32$) New World to give $3x - 24$ with disrupted meiosis. On doubling, this yielded a more fertile hexaploid which, through selection, give individuals with yet higher fertility in later generations. A natural hexaploid, probably not closely related, *A. montana* is known from Manchuria, (the provinces collectively of Liaoning, Jilin and Heilongkiang as the "North East" of China).

DNA and the Modern Interest

Against this more classical background it is appropriate to consider the contemporary interest in *Anemone*.

Hantula et al (1989) examined *Anemone* in the broad sense, that is including representatives of *Hepatica* and *Pulsatilla* using cp. DNA variation. Their conclusion was that the three entities should be treated on the same level either as separate genera or subgenera of *Anemone*. cp. DNA data here is not decisive but merely adds to what taxonomists already have. Hoot and co-workers in a series of papers (Hoot and Palmer 1994, Hoot et al 1994 and Hoot 1995) have clarified matters along the following lines. Hoot and Palmer (1994) showed for cp. DNA parallel inversions within these related genera. Hoot et al 1994 and combining morphological and cp. DNA, opted to regard *Hepatica* and *Pulsatilla* and groups (sections) within *Anemone*. Finally, Hoot (1995) with reference to *Anemone* showed that species based on $x = 7$ and $x = 8$ tended to cluster separately. Of the Chinese species included in the enquiry, *A. demissa, flaccida, huphensis narcissiflora, obtusiloba, tomentosa, rivularis, rupestris* and *vitifolia* tended to cluster much as set out in Table 11.1. An additional species not shown here *A. dichotoma* (Sect. Anemonidium) maintained an appropriate separation. One, perhaps surprising, exception was the clustering of *A. rivularis* (sect. Rivularidium) with *A. rupicola* (sect. Anemone).

With reference to the relation of species occurring in different hemispheres, it is interesting that Hoot (1995) showed the cp. DNA of *A. dichotoma* and *A. canadensis* to share two unique inversions.

A Retrospective

Would we be justified in drawing the conclusion that as an old relatively primitive angiosperm *Anemone* was widely distributed relatively early and, given the rise of the Himalayas, responded only modestly to the new opportunities provided? *Clematis* in the same family has, on this basis, done rather more in the way of speciation but nothing like as prolifically as the gamopetalous genus *Primula* in the Metachlamydeae, arguably, more recently differentiated than the Archichlamydeae.

On this reckoning, are we to regard *Camellia*, for whatever reason, as having only recently begun to speciate? Conversely, was the Ericaceae with its rhododendroid nuclear population, fortuitously with everything in place, ready and able to move into evolutionary overdrive with the rise of the Himalayas? Before considering *Rhododendron* we pause to examine *Primula*.

Primula

Of some 500 species worldwide about 300 occur in China, of which perhaps 206 are described as endemic, occur in China and nowhere else naturally. It is, though, necessary to inject a note of common sense into the use of the term endemic in a Chinese context.

Political boundaries can be relatively arbitrary and take little account of ecology. A *Primula* species, or indeed any other species, might be considered endemic until it were found over the border in so small an area, for example, as Sikkim. At this point its endemic status vanishes. The term endemic is rather more use if it is applied to a particular kind of habitat. "Endemic to the Himalayas" might make more botanical sense regardless of several political boundaries being crossed. Even so, political or administrative boundaries can have their uses as the following table shows.

Table 11. 4. A view of *Primula* distribution in China

Total species occurring in China		300
Species 'endemic' to China of which		206
endemic to	Sichuan	41
" "	Xizang (Tibet)	35
" "	Yunnan	39
" "	Sichuan + Xizang	0
" "	" + Yunnan	33
" "	Xizang + Yunnan	6
" "	Sichuan + Xizang + Yunnan	12
" "	other provinces combined	40

(166 — Sichuan, Xizang (Tibet), Yunnan, Sichuan + Xizang, " + Yunnan, Xizang + Yunnan, Sichuan + Xizang + Yunnan grouped)

Some conclusions can be drawn at once, namely that the three provinces are rich in *Primula* species, that endemics shared between Xijang and Sichuan or between Xizang and Yunnan are relatively rare, whereas Sichuan and Yunnan between them have more in common, implying that Tibetan *Primula* spp. are somehow "different". While, therefore, a *Primula* enthusiast might want to concentrate on three obviously species-rich provinces, it would be unwise to ignore other provinces. *P. obconica, farinosa* and *malacoides* are all more widely spread in China. That said, it has to be admitted that such esteemed species as *P. bulleyana* and *florindae* originate from the selected areas highlighted in Table 11.4.

Primula Subdivided

While it is open to any botanist to subdivide a genus (into sections, subgenera or even separate genera), such a scheme is only likely to be adopted if it has obvious advantages. Several schemes have been proposed for *Primula*, of which that adopted here is used in the *Flora of China*, (Wu and Raven 1996), where the genus is divided into 24 sections (see Plate 12). The obvious implication is that species within a section share more features than do those in different sections, but rather less obvious is how any section stands in relation to the others. Is it, for example, an "advanced" group or a "primitive" one? Is it at the base of the genus so that other sections derive from it or is it some peculiar outlier, which has evolved to meet special circumstances? More particularly in the Chinese context, are any of the sections regionally confined? As Table 11.5 shows, not all the sections are of equal size, and this can be significant.

The distribution of the various sections can be described as follows. Except for section Ranunculoides (found in Anhui, Hubei, Hunan, Jiangxi and Zheijang and consisting of only two species), all other sections are represented in Sichuan or Xizang or Yunnan or more than one of these, however else they may be spread across China. In one or two sections their entire representation is confined to these provinces – for example Malvacea, is in Sichuan and Yunnan, Picnoloba (only one species) to Sichuan, Primula (one species)Xizang, Bullatae (four species), Xizang, Souliei (six species) Sichuan and Yunnan. Some other sections that have representation beyond China are best regarded anyway as Himalayan notably sections Amethestina, Sikkimensis, Cordifoliae, Minutissima, Dryadifolia and Capitatae.

Table 11. 5. *Primula* sections within China

Section	No of species	Sample chromosome nos.
Monocarpicae	13	18, 22, 24, 36, 66
Obconicolisteri	10	22, 24, 62
Cortusoides	17	22, 24, 29 36
Malvacea	7	-
Auganthus	3	22, 24, 36, 48
Pycnoloba	1	24
Ranunculoides	2	-
Primula	1	22
Carolinella	8	-
Bullatae	4	24
Petiolares	49	22, 36
Proliferae	19	22 (mostly), 24
Amethystina	8	-
Sikkimensis	8	18, 22 (mostly)
Crystallophlomis	41	18, 22, 22 [a], 44[a]
Cordifoliae	5	22
Aleuritia	46	16, 18, 20, 22, 32, 34, 36, 44, 72
Minutissima	17	18, 22
Souliei	6	16
Dryadifolia	3	-
Denticulata	6	22, 32, 44 [a]
Capitatae	2	18, 44 [a]
Muscarioides	13	20, 40 [a]
Soldanelloides	11	20
	300	[a] Plus 1 - several supernumeries sometimes.

Note :   Not all species have been counted for chromosomes in some sections.

The overall pattern which emerges is of a genus diverse throughout much of China but brought to a crescendo of variation in Sichuan, Xizang and Yunnan. This situation described here at some length for *Primula* finds an echo in many other Chinese genera so much so that one might coin the phrase *Primula*-type distribution.

An Aside on Variation in *Primula*

We may gauge the age of many genera by discoveries from palynology the study of pollen. Muller (1970) using evidence from fossil pollen showed that many angiosperm taxa predate the Miocene – a time of considerable upthrust. The assumption is that *Primula* and many other genera widely spread in China were well placed to exploit the many new habitats created by Himalayan uplift. Although species can be of relatively recent evolution their genera in many cases predate the world's mountain systems upon which they have diversified, a point raised earlier.

The response of *Primula* to the appearance of new habitats prompts comparison with *Anemone*. One might argue that *Anemone* is the appreciably older genus of the two and, for reasons we do not understand, had genetically settled down. Has perhaps its response been more accommodation than exploitation hence the comparatively meagre diversity apparent?

Chromosome Evolution in *Primula*

A recognised trend in chromosome evolution is a stepwise reduction in chromosome number. The classic study of *Crepis* (Compositae) inaugurated by Babcock (1942) showed a change from long–lived perennials to short–lived annuals accompanied by increased seed production and a simplified plant body. In present–day language we could describe loss of chromosome material as shedding of unwanted DNA. It should be recognised that not all reduction in chromosome number operates in a similar fashion. If, for example, two chromosomes with terminal centromere fuse to create a single metacentric chromosome (a so-called Robertsonian fusion), the amount of DNA present remains virtually unchanged.

Suppose stepwise reduction in chromosome number leads to the evolution of plants relatively highly adapted to a restricted range of habitats. If, then, two fairly closely related species resulting from similar processes hybridise, one can envisage a plant encompassing the eco-adaptedness of both parents. Next if chromosome number doubles to create a stable allopolyploid, there is the more or less instant evolution of a new species.

Fig. 11. 2.

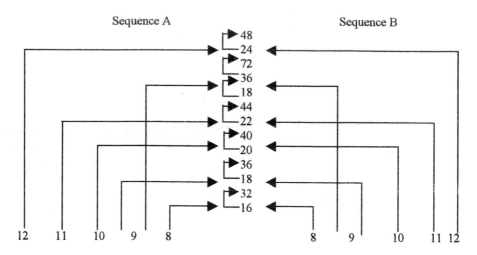

In the context of *Primula* cytogenetics (and concentrating on the gamete number) the following sequences of events could have been entirely possible. Sequences A and B each imply 12 → 11 → 10 → 9 → 8 reductions of basic number. Hybridisation between appropriate gamete numbers and then, with or without doubling, yields some of the somatic counts given in Table 11.5. Some somatic numbers such as 29, 62 and 66 for example are less easily resolved.

The most famous and well documented instance is that involving non-Chinese species where *P. floribunda* (2*n* = 18) *x P. verticillata* (2*n* = 18) gave *P.* x *kewensis* 2*n* = 18 a relatively sterile diploid *P* x *kewensis* which, on doubling spontaneously, gave a vigorous fertile species *P.* x *kewensis* 2n = 36.

We now turn to the remarkable case of *Rhododendron* and its proliferation of species in China.

12 *Rhododendron*

The genus *Rhododendron* comprises about 900 species of which many are native to China, the remainder being spread more thinly across the northern hemisphere. Since so many of them are attractive garden subjects it is hardly surprising that plant hunters have returned repeatedly to natural concentrations of diversity, seeking material either already suitable for gardens or perhaps useful for breeding. To understand the genus it is necessary to know something of its classification.

Subdividing the Genus

Rhododendron in its most familiar form is an evergreen shrub with smooth leaves and is represented by such famous varieties as Lord Roberts, Pink Pearl and Sappho. Many gardeners would regard azaleas as appreciably different to *Rhododendron* – perhaps so different as to be put in a separate genus. "*Azalea mollis*" is a common item on sale in garden centres. Even so, the balance of botanical opinion is to include azaleas within *Rhododendron*. There is, however, far more to this genus than this opening discussion might indicate. As a basis for what follows the summary of a modern classification by Cullen et al. (1997) is shown in Table 12.1. Plates 12, 13 and 14 illustrate various parts of the genus.

At this stage it is necessary to recognize that some terms, though useful, are informal. Thus we distinguish between lepidote and elepidote plants. The former is recognised by scales which occur on almost all parts of the plant. Elepidote types can be split in which the indumentum when present is at least partly composed of compound hairs (sub genus Hymenanthes) and those where the hairs are always simple. "Azalea" again is useful and one should recognise it can include both deciduous (subgenus Pentanthera of which "*Azalea mollis*" is an example and subgenus *Tsutsusi* which includes the Belgian pot azaleas' and the Japanese Kurume azaleas), and evergreen plants.

The Geographical Origins of Garden *Rhododendron*

Due to the efforts of dedicated plant collectors as outlined in Chapter 4 and others, a large number of these species, many of them from China, embellish temperate gardens around the world in both hemispheres. Using Cullen (1997) as a basis, Table 12.2 indicates something of their origins.

Table 12. 1. A summary of *Rhododendron* classification. (After Cullen 1997)

	Subgenus Rhododendron		
	Section	Pogonanthrum	
	Section	Rhododendron	
"lepidote"		Subsections	Edgeworthia, Virgata, Scabrifolia, Cinnabarina, Triflora, Rhodorastra, Lapponica, Saluenensia, Maddenia, Glauca, Micrantha, Monantha, Rhododendron, Tephropepla, Moupinensia, Uniflora, Heliolepida, Caroliniana, Camelliiflora, Afghanica, Campylogyna, Boothia, Trichoclada, Baileya, Genesteriana, Lepidota
	Section	Vireya	
		Subsections	Pseudovireya, Siphonovireya, Phaeovireya, Malayovireya, Albovireya, Solenovireya, Euvireya
non or e-lepidote	Subgenus Hymenanthes		
	Section	Pontica	
		Subsections	Auriculata, Griersoniana, Glischra, Williamsiana, Selensia, Parishia, Neriiflora, Maculifera, Venatora, Barbata, Grandia, Fortunea, Falconera, Taliensia, Irrorata, Campylocarpa, Fulva, Argyrophylla, Lanata, Arborea, Fulgensia, Thomsonia, Pontica, Campanulata
Azalea complex	Subgenus Tsutsusi		
	Section	Tsutsusi (evergreen azaleas)	
	Section	Brachycalyx	
	Subgenus Pentanthera		
	Section	Pentanthera (cult. decid. Azaleas)	
	Section	Viscidula	
	Section	Rhodora	
	Section	Sciadorhodion	
	Subgenus Therorhodion		
	Subgenus Azaleastrum		
	Section	Azaleastrum (evergreen)	
	Section	Choniastrum (evergreen)	
	Subgenus Mumeazalea		
	Subgenus Candidastrum		

Note: Rhododendron is both a 'common name' and the formal name of the genus where it will appear in italic. Where it is used as a subgenus, section or subsection name it will be non-italicised. Provided the reader is aware of the problem it is normally apparent from the context what is intended.

Clearly, the contents of Table 12.2 are likely to be biased to what seemed worth collecting in its original location and subsequently perpetuating through the efforts of nurserymen. One should recognise, too, that in terms of available diversity, the efforts of breeders have yielded an immense number of varieties, hundreds of them well worth growing as appealing ornamentals of great beauty.

From Table 12.2, three points are worth stressing.

1. 85 species originate from plants indigenous to Sichuan, Xizang or Yunnan or shared among them.
2. Of the 82 species from China and beyond, Sichuan, Xizang and Yunnan have also contributed material.
3. Of the species referred to here less than 20 originate from the New World. Were we to consider what sample of species interests American horticulturists (provided growing conditions are broadly comparable) it would not differ appreciably from the European counterpart.

(For a recent contribution to Rhododendron taxonomy see Kurashige et al 2000 and upon that is a brief comment subsequently.)

Table 12. 2. Geographical origins of 291 *Rhododendron* species in cultivation in European gardens

Species occurring in:	
Sichuan	23
Xizang	11
Yunnan	18
Sichuan + Yunnan	22
Sichuan + Xizang + Yunnan	11
Sichuan + Xizang + Yunnan + other Chinese provinces	20
Chinese provinces excluding Sichuan + Xizang + Yunnan	4
Taiwan	4
China and beyond	82
Himalaya	16
Old World, non Himalayan	57
New World	18
Garden origin, unknown in wild	5

Note: For simplicity varieties and sub species have been disregarded.

It is, therefore, not difficult to appreciate the importance, numerically, of these three Chinese provinces for *Rhododendron*. Beyond this, is it possible to specify the situation more precisely? One approach, confined for the purpose of illustration to the 52 species, in the first three rows, that is, indigenous to Sichuan or Xizang or Yunnan neglecting those more widely spread, even where they are shared between two of these provinces might be as follows. Suppose, too, that these 52 species are considered using the arrangement set out in Table 12.1. We can then assemble Table 12.3.

Table 12.3. Taxonomic diversity among *Rhododendron* species endemic to Sichuan or Xizang or Yunnan

Combined species total	Subgenera represented	Sections represented	Si	Xi	Yu	
	Rhododendron	Pogonathum	1	2	0	3
52		Rhododendron	12	5	11	28
	Hymenanthes	Ponticum	10	4	7	21
			23	11	18	52

Not only therefore have the three provinces, on this basis provided a significant number of species they have also made available a *diversity* of species. Note that for *sections* Rhododendron and Ponticum there is among them in these three provinces a fair representation of their *sub*sections.

It will be obvious to the reader that attention has been given only to relatively confined species. Were we to include species common to all three provinces and beyond but still within China, the available diversity would increase. The advantage of combining Tables 12.1 and 12.2 is to introduce a degree of precision. How far matters are taken further depends on the level of the reader's own interest in that direction.

An Ecological Note

Although many rhododendrons in cultivation tolerate the indifferent soils to which unthinking gardeners commit them, a consistent theme in the ecology of this genus is its preference for acid peaty substrates. This, in turn, has helped create, to the despair of ecologists, a mass market for peat, virtually a finite resource. Surprisingly perhaps, some Rhododendrons are adapted in nature to habitats on limestone.

To most people this genus is one rooted in soil and unfamiliar as an epiphyte. Beyond China in New Guinea a minute plant, *R. caespitosum*, grows as an epiphyte, for example, on tree ferns, *Cyathea tomentosa*. The dimensions of this Rhododendron are stems about 2.0 cm tall, leaves 6 times 3 mm and pink flowers up to 12 mm across. It belongs to section Vireya (Argent et al. 1999).

Hybridisation and New Varieties: a Rationale

From the hands of breeders a large number of new hybrids become available to gardeners each year. At a more technical level, to probe the available information is to discover a network of interrelated pedigrees that calls for some rational guide for the serious enquirer. What follows here, therefore, is a simplified (and admittedly botanical) view of *Rhododendron* hybridisation. No attempt is made to follow the intricacies of pedigree nor to dilate upon the merits of this or that variety. Emphasis is upon the basic principles involved.

With reference to Table 12.1 we can make the following initial assumptions.

1. Subgenera are only distantly related and therefore unlikely to hybridise.
2. Within a subgenus, relations between sections will be closer and hybridisation perhaps more likely though seldom commonplace in nature and requiring human intervention in culture.
3. Where subsections exist hybrids between them could reasonably be expected.
4. Within a subsection plants are relatively similar, by definition, and it might be surprising if they did *not* form hybrids, even in the wild, unaided by ourselves.

With this approach, the reader is offered Table 12.4 to indicate what is not more than an informative sample of crosses.

There are some surprises, the most obvious one being that while it is just possible to secure a very few inter-subgenus crosses, at the next level down, crosses within a subgenus between sections furnish no examples (Chamberlain, pers. comm.). Another is that within the Azalea complex subgenus Pentanthera, section Sciadorhodion are four species *R. albrechtii, pentaphylhum, schlippenbachii* and *quinquefolium*, which more recent opinion considers widely dissimilar (Chamberlain, pers. comm.). This clearly raises an interesting question over the absence of section by section crosses. It suggests that what have been called sections might need to be raised to subgenus level to reflect the evolutionary divergences suggested both by failure to produce hybrids and emerging DNA evidence. As yet, there is not a large body of such evidence upon which to go and the reader is directed to Kurashige et al. (2000), where a polyphyletic basis is suggested for subgenera Pentanthera and Azaleastrum, for example.

A Further Consideration of *Rhododendron* Breeding

Despite the earlier indicated intention to avoid the minutiae of pedigree, the foregoing material does permit a consideration of the three varieties used to introduce the genus.

1. Lord Roberts. This is an old variety reckoned to derive from *R. catawbiense* although what else was added remains uncertain.
2. Pink Pearl. A hybrid *R. griffithianum* x *catawbiense* yielded George Hardy. This crossed with Broughtonii, the origin of which is apparently obscure, yielded Pink Pearl.
3. 'Sappho'. This variety arose from *R. ponticum* times some unknown other parent.

All these varieties fall with subgenus Hymenanthes, section Pontica and all originated toward the close of the 19th century. That same section a century later

Table 12. 4. A summary of hybridisation in the genus *Rhododendron*

Status	Examples		Comment	Reference
	Hymenanthes (sect. Pontica) R. Griersonium	× Rhododendron (sect. Rhododendron) × R.	An apparently unique non-lepidote × lepidote hybrid yielding var. 'Grier-dal'	Rouse & Williams (1988)
Subgenus × subgenus	Rhododendron (sect. Vireya)	× Pentanthera (sect. Pentanthera)	Flowers deformed and sterile	"
	Hymenanthes (sect. Pontica) R. catawbiense	× Pentanthera (sect. Pentanthera) × R. viscosum	The origin of 'Azaleodendron'	Moser (1991)
within a subgenus section × section			No authentic examples are known.	
Within subgenus and Section subsection × subsection	Hymenanthes (sect Pontica) Auriculata R. auriculatum Arborea R. arboreum	× Fortunea × decorum × Thomsoniana × R. thomsonii		van Gelderen and van Hoey Smith (1992)
Within a subgenus section and sub section within	Hymenanthes (sect Pontica) Fortunea R. griffithianum Pontica	× Fortunea × R. fortunei	origin of var Angelo	"
Subsection crossing	Pontica R. caucasicum Pontica	× R. catawbiense	origin of Boule de Neige	"
	R. ponticum	× R. ponticum (?)	origin of Blue Peter	"

Increasing Likelihood of Hybridisation →

Diminishing Closeness of Relationship ←

Table 12. 5. Significant species in *Rhododendron* Breeding

Section Pontica

sub section			
	Auriculata	*R. auriculatum*	
	Griersoniana	*R. griersonianum*	Yunnan, Burma
	Williamsiana	*R williamsii*	Sichuan
	Parishia	*R facetum*	Yunnan, Burma
	Neriiflora	*R dichroanthum*	Yunnan, Burma
		R forrestii	Xizang, Yunnan, Burma
		R haematodes	" " "
	Fortunea	*R fortunei*	China
		R orbiculare	Sichuan
	Campylocarpa	*R campylocar-* *pum*	Himalaya, Sichuan, Yunnan, Burma
		R wardii	Sichuan, Xizang, Yunnan
	Arborea	*R arboreum*	Himalaya, India, China
	Thomsoniana	*R thomsonii*	Himalaya, Xizang
	Pontica	*R catawbiense*	north America
		R caucasicum	Europe
		R yakushimanum	Japan
	Campanulata	*R. campanulatum*	

still monopolises the efforts of breeders and, given the lovely varieties they continue to produce, the store still seems far from exhausted. A retrospective is perhaps appropriate here. Several species have contributed conspicuously to the development of modern varieties. By subsection, each is listed together with its geographical distribution in Table 12.5. (The order corresponds to that in Table 12.1).

Perhaps a question, which presents itself almost unasked to the reader, is what might have happened to *Rhododendron* breeding had intrepid botanists like Kingdon Ward not braved conditions in Sichuan, Xizang and Yunnan? That said, it is evident that besides this there has been a more widespread contribution from across the Northern Hemisphere.

It is worth pointing out in conclusion that not all the effort at *Rhododendron* improvement has been directed at the section Pontica. The Japanese have for several hundred years been generating and selecting hybrids within subgenus Tsutsusi. The subgenus Pentanthera has supplied all the Mollis, Rustica, Ghent and Exbury "Azaleas", and the subgenus Rhododendron has supplied dwarf rock-garden hybrids.

For the reader primarily concerned with Rhododendrons as horticultural material, helpful references are Cox and Cox (1997), van Gelderen and van Hoey Smith (1992), Moser (1991), Postan (1996) and Salley and Greer (1986).

Rhododendron Success in China

How might the remarkable diversity of this genus in China be explained? The short answer is that we do no know. Two things, however, stand out and are probably significant. The first from table 12.4 is that as one ascends the taxonomic hierarchy frequency of crossing (as expected) tends to decrease. The second, apparent from breeders' efforts, is the astonishing diversity they have been able to unlock in the form of new varieties. Might we assume that the possession of such intrinsic variability in nature allowed speciation on so impressive a scale? There is, of course, an alternative possibility: that the changing environment of western China *promoted* the generation of variability.

Perhaps it is appropriate here to leave the reader with a comment by Darwin (1868):

"The most celebrated horticulturist in France, namely, Vilmorin, even maintains that, when any particular variation is desired, the first step is to get the plant to vary in any manner whatever, and to go on selecting the most variable individuals, even though they vary in the wrong direction; for the fixed character of the species being once broken, the desired variation will sooner or later appear".

Could it be that what was Vilmorin's experience happened also in nature, to *Rhododendron*, given the challenge and opportunity of so many newly arising habitats?

13 *Rosa*

This chapter is included to provide an instructive contrast to those dealing with, for example, *Clematis, Gentiana, Primula* and *Rhododendron*. In those cases, China has provided, for each, large numbers of species which can, in many cases, be taken into Western cultivation with little or no genetic modification at the hands of breeders. The story for Chinese roses is different. Here, relatively few species were involved but which, early on, entered the main stream of Western rose improvement. While, as will be shown, repeat flowering was a major Chinese contribution, it was not the only one. We do, however, need to recognise how, from early in the nineteenth century, it was outside of China that its roses were drawn into the sharply focussed objectives of commercial rose breeding.

Botanical and Horticultural Viewpoints

To open any reasonably comprehensive rose catalogue is to be confronted by the stock in trade of Damask, Gallica, Bourbon, Hybrid Tea, Floribunda and Polyantha and some others which might be mentioned. Rose specialists are, of course, at home with all this. The knowing gardener has at least some idea of where his or her preferences lie, and the botanist, much as he likes roses and can admire what breeders have done, is likely to be uncomfortable with the categories. It is as if horticulturists view roses with a pair of spectacles quite different to those a botanist might wear. Neither view has pre-eminence, it is quite simply that each discipline has different priorities and in this chapter, using mostly Chinese rose material for illustration, an attempt is made to see these plants in both ways.

Table 13.1, for all its likely unfamiliarity, is a botanist's view of the genus *Rosa*. Its starting point is the worldwide existence of *wild* roses and it provides the foundation for several things. It is essential to ecology. Not every rose grows everywhere and each has the ability to grow better in one place than another. Genetically it can be shown that some roses are derived from others and it becomes possible to see an evolutionary pattern of "earlier" and "later" types. Rose breeding has in turn generated some of the insights which enable botanists to improve their own ideas about what constitutes a realistic classification of wild species. What makes the roses so interesting is that to a far greater extent than for *Gentiana, Primula* or *Rhododendron* they have, in some cases, been transformed.

Table 13. 1. A taxonomic summary of the genus *Rosa*

(Chromosome numbers per cell in parentheses)

		Distribution	Examples of species
Subgenus	Hulthemia	Iran, Turkey	*R. persica* (14)
Subgenus	Hesperhodos	Western N. America	*R. stellata* (14)
Subgenus	Platyrhodon	China, Vietnam	*R. roxburghii* (14)
Sungenus	Rosa		
Section	Gallicanae	Europe, Near East	*R. centifolia* (28), *damascena* (28), *gallica* (28)
	Caninae	Europe, Near East, Iran	*R. canina* (35), *eglanteria* (35), *tomentosa* (35)
	Carolinae	USA and southern Canada	*R. carolina* (28), *foliolosa, nitida* (14), *palustris* (14), *virginiana*
	Pimpinellifoliae	Europe, Near East China (Western group)	*R. pimpinellifolia* (28), *R. foetida* (28), *dunwichensis* (28)
		(Eastern group)	*R. ecae* (14), *hugonis* (14), *koreana* (14), *sericea* (14), *xanthina* (14)
	Cinnamomeae	N. Hemisphere except Western Europe	*R. beggeriana* (14) *R. kordesii, moyesii* (42), *nukana* (42), *rugosa* (14), *webbiana* (14)
	Chinensis	China, Vietnam, Himalayas	*R. chinensis* (14, 21, 28), *gigantea* (14) *odorata* (14), *x borboniana*
	Synstylae	Eastern N. America, Temperate Eurasia	*R. filipes* (14), *moschata* (14), *mulligani* (14), *multiflora* (14), *wichuriana* (14)
	Banksianae	China, Himalayas, Vietnam	*R. banksiae* (14), *R. cymosa* (14)
	Laevigatae	East China	*R. laevigata* (14)
	Bracteatae	China, USA, Vietnam	*R.bracteata* (14)

almost out of recognition from their wild ancestors. An obvious example to take is the hybrid tea rose. To a botanist it is obvious that backward-pointing thorns conveniently help a briar to scramble over or between competing vegetation. Having done so, buds form and flowers open to reveal a mass of anthers shedding pollen and stigmas available for cross-fertilisation. The hybrid tea rose growing without competition as a small *bush* in the middle of a bed, pruned annually to little more that a stump, most of its anthers converted to petals and its thorns a glaring atavism is, on this reckoning, quite simply, an absurdity. If we replace our botanical view with a horticultural one we recognise largely different priorities. For the present, the hybrid tea remains the focus of most commercial rose growing. Here and there are signs, though, of a reappraisal, a part of which is for breeders to recover earlier and more natural forms of rose growth habit. An obvious example is the New English line developed by David Austin. There is, too, a sustained, if minority, interest in growing purportedly wild species of rose, even if, in many cases, these are striking or useful variants. Against such a perhaps surprising introduction one can better consider the remarkable achievements of more than 150 years of rose breeding. One particular aim here is to describe the extent to which roses of Chinese origin have contributed to those enjoyed around the world. Another, of course, is to link botanical and horticultural perceptions of rose breeding

Table 13.1 shows the principal subdivisions of the genus *Rosa*, their approximate geographical range and some of the species, together with their chromosome numbers, which have contributed to commercial rose growing, and hybridisation (see also Plate 15). For a treatment of roses in China see Yu (1985).

Rose Culture in Chinese History

Ogisu (1996) has provided an accessible treatment of roses in China's traditional horticulture. Records of rose growing are known from the sixth century A.D. and from the Song and Ming Dynasties. The Koushin rose, apparently recurrent flowering, had arrived in Japan by either the Sui or Tang dynasties, more than 1000 years ago. Such was the interest among the Chinese that by the time the Europeans arrived, numerous varieties were available which differed from their wild ancestors. Among those introduced to Europe, four are perceived to be of importance. Their dates of introduction to Europe are provided by Hurst (1941). They are from *R. chinensis*, Slater's Crimson China (1792) and from *R. chinensis* x *R. gigantea*, Parson's Pink China (1793), Hume's Blush tea scented China (1809), and Park's Yellow tea scented China (1824). The last of these is now reckoned extinct. *R. chinensis*, the wild ancestor, has proved to be somewhat elusive but Ogisu (1996), has provided evidence of its occurrence in Sichuan.

Repeat Flowering: a Modern Perception

Repeat or recurrent flowering arose as a mutant in *R. chinensis* and was used in rose improvement as early as the Song dynasty. Recently, the physiological basis of repeat flowering has been examined by Roberts et al. (1999). Shoot growth, vegetatively, is terminated by flowering. Further growth is then axillary. In short season roses such growth is not then terminated by flowering. Conversely, in recurrent roses, flowering repeatedly terminates such growth and flowers appear throughout the season. When levels of gibberellins (GA_1 and GA_3) are examined they are found to be lower, and to remain lower, in recurrent flowering plants. (Earlier work by Zieslin and Halevy in 1976 had shown that exogenously applied GA_3 could suppress flowering in a repeat flowering rose, Baccara if it were supplied in sufficiently high concentrations).

A feature, therefore, deriving from China roses and, for a century and a half integral to rose breeding in the West, is finally understood in terms of plant physiology. Roberts et al. (1999) remark that, in terms of short season and repeat flowering, well contrasted types offer an opportunity for a molecular approach to this subject.

China Roses in the West

Using *R. moschata* (Synstylae) with Parson's Pink and later Parks' Yellow Tea-Scented (Chinensis) represent intersection crosses which yielded Noisettes and eventually yellow Teas – a development involving only diploid roses. A related development was that of crossing *R. gallica* (Gallicanae) with Hume's Blush Tea-Scented (Chinensis) a tetraploid x diploid to provide a new and triploid group the Hybrid Chinas only (as seen, in retrospect, from the 20th century) when the Hybrid Chinas mutated to yield the Hybrid Perpetual in about 1837.
Persistence pays sometimes and rose breeders repeatedly crossed Tea roses (diploid) with Hybrid Perpetuals (tetraploid) to yield a succession of Hybrid Teas – normally, as would be expected, sterile triploids. Eventually tetraploid forms occurred (though, of course, not so perceived at the time) and the way was opened for the diversification of Hybrid Teas by incorporating material from other tetraploids, notably *R. foetida* (Pimpinellifoliae), a source of strong yellow flower colour.

The Oriental contribution to rose breeding was by no means exhausted at this point. Certain species of Synstylae occur in China, notably *R. multiflora* and *R. wichuriana* (this latter from Japan) and contributed to the familiar rambler, Albertine, for example. *R. bracteata* (Bracteatae) hybridised with a Tea Rose (the exact variety is unknown) but of Chinese ancestry and yielded the remarkable variety Mermaid, which, though diploid has only sparingly contributed striking descendants.

Today, we have available a wider choice of roses than at any time previously. The bequest from China of prolonged flowering has now entered virtually every kind of modern hybrid rose type and is an enduring contribution.

Such old roses as Charles de Mills, with its remarkable colour and flower conformation, have a place, but anyone who has grown it recognises both how short is its blooming season and the need to do something about it hence the evident value to the breeder of Oriental roses.

The literature on roses, both those from China and elsewhere, is vast and rather than overburden the reader, one section, Pimpinellifoliae, represented both within and beyond China is selected for further comment.

Section Pimpinellifoliae

Roses belonging to this section are distinctive having small leaflets and, typically, yellow, cream or white flowers and are found in various parts of the Northern Hemisphere in the Old World. Taxonomic opinion is not uniform but the cytogenetic work of Roberts (1977) provides a helpful starting point. See Table 13.2. Part of the interest here is in *R. foetida* of Iranian origin, reckoned to be the source of strong yellow among hybrid tea roses.

Table 13. 2. Ploidy and simplified distribution of *Rosa*, section Pimpinellifoliae.
(After Roberts 1977)

Tetraploids	*R. spinosissima*	*R. hemispherica*	*R. foetida*
	(Europe, Turkey, USSR, Mongolia, China)	(Turkey, Iran)	(Turkey, Syria, Iran, Afghanistan, USSR, India, Pakistan, Kashmir, China, Mongolia, Korea)
Diploids	*R. primula* (Turkestan)	*R. ecae* (Afghanistan, Pakistan, Turkestan, China (Shansi)).	*R. sericea* (India, Nepal, Durma, China)
	R. hugonis/xanthina (China, Mongolia, Korea)	*R. omiensis* (China, India, Assam, Simkkim)	*R. forrestiana* (China)
	koreana (Korea)	*farreri* (Known only in cultivation)	

The essentials of Roberts (1977) approach based on some 36 taxonomic discriminants, are that on morphology, three species *R. forrestiana, koreana* and *farreri* belong more appropriately among Cinnamomiae and that *R. hugonis* and *xanthina* perhaps should be regarded as a single species, so similar are they. Beyond this, hybridisation data is not extensive but is sufficient to show high viability and the seeming absence of sterility barriers among those retained within Pimpinellifoliae. Both diploid *R. foetida* and tetraploid *R. spinosissima* have been involved in relatively wider crosses in the development of modern rose varieties. The latter, sometimes known as *R. pimpinellifolia*, has contributed to such famous varieties as Frülingsgold, Frülingsmorgen and Maigold.

The scheme proposed by Roberts, shown in table 13.2, has more recently been examined by Haw (1996). The points of interest to emerge are this author's opinion reinstating *R. hugonis* as a species separate from *R. xanthina* and also differing from Roberts in not regarding *R. omiensis* as a variety of *R. sericea*. (For a comprehensive account of the way in which this and the other sections of the subgenus Rosa have contributed to rose breeding see, for example, Beales 1997.)

Named Varieties and Taxonomic Change

Unless a variety is known to be the result of deliberate crossing by a breeder we can only *infer* the species involved. If, in implicating a particular wild species, it is found that it has been subject to taxonomic revision, confusion can arise. Two (or three) species within Pimpinellifoliae serve to illustrate this situation with the aid of some well known cultivars. Suppose, initially, we regard *R. hugonis* as a form of *R. xanthina*. The variety Canary Bird on this basis arises from within *R.xanthina* the view taken by Philips and Rix (1993). If however they are regarded as separate species (Beales 1997) then Canary Bird is allegedly an interspecific hybrid *R. hugonis* x *R. xanthina*. (Unless seeds were harvested from one of these parents and grown on, not only the parentage but the direction of cross is also an inference.)

A similar problem arises with a rose which, for simplicity, will be referred to initially as Cantabrigiensis. This very attractive clear yellow single rose originated as a seedling in the Cambridge University botanic garden and was recognised in 1931. On the basis of its morphology its putative parents were thought to be *R. hugonis* and *R. sericea* so that it could be referred to as *R* x *cantabrigiensis*. Austin (1988) refers to it as *R. pteragonis* Cantabrigiensis. Matters fall into place if we regard *R, pteragonis* as itself a hybrid, *R. hugonis* x *R. sericea* and refer to it more informatively as *R* x *pteragonis*.

That said, it should be recognised that, despite the shifting opinions as to the names to be employed, expert opinion has little difficulty in recognising the *kind* of germ plasm involved in the origins of Canary Bird and *R* x *cantabrigiensis*.

Stage in Rose Improvement

Historically, one can identify three major phases in rose improvement. The first phase involved merely the perpetuation of striking variants such as doubleness or, colour change or, in the case of *R. centifolia* the mutant condition for 'mossiness', a conspicuous fuzzy growth on the sepals. Following on from this there is a stage where, given increasing familiarity with the genus, chance hybrids or purported hybrids are perpetuated. Hybrids between *R. chinensis* and *R. gigantea* would fall in this category as would *R.* x *cantabrigiensis* referred to above. Finally there is the present sophisticated situation where natural relationships within the genus are well understood, chromosome number is taken into account and breeders are working toward well-defined objectives. Clearly, the phases overlap, and even today the simple perpetuation of some novelty is part of the commercial scene. The Austin variety Mary Rose (1983) yielded the white variant Winchester Cathedral introduced in 1988, for example.

A significant feature of rose improvement is that of retrospective awareness. Various developments made in the 19th century created new and appealing kinds of rose. What had been achieved in botanical terms was only appreciated perhaps a century or more after. The following is an example quoted by Wylie (1954).

On Reunion (known as L'Ile de Bourbon) two roses, a damask, Rose of the Four Seasons and Parson's Pink China were used as field hedges. In 1817 a suspected hybrid was found and progeny raised from it yielding a variety Rosier de L'Ile de Bourbon but not shown to be a triploid, from having European tetraploid and Oriental diploid parentage until the 20th century.

Most wild roses have a relatively straggly growth habit between and around competing vegetation. The breeder often seeks to incorporate or select for compact growth. It is hardly surprising if occasionally a subtle change in constitution then turns otherwise recognisable varieties from bush back to something more ancestral. This can be used to advantage. The old hybrid tea variety Ena Harkness tends to droop its flowers. Conveniently, this is usefully shown off in the climbing variant, perhaps currently, the commoner form of this variety. For discussion of climbing mutants see Wylie (1955). The matter remains somewhat speculative.

Alongside of modern rose breeding there remains a minority interest in wild roses as garden subjects although, as pointed out earlier, those more widely grown are in most cases slightly different from their genuinely commonplace relatives. From China the following are of horticultural interest. (See Plate 15)

A familiar rose nowadays is *R. moyesii* (Cinnamomae). It is a hexaploid species with 42 chromosomes. While of little proven use as a breeding parent, it exists in several distinctive variants grown today in temperate gardens. Others from China include *R. sericea pteracantha* (Pimpinellifoliae) with its spectacular thorns, and the several variants of *R. banksiae* (Banksianae). Another is *R. laevigata* (best known through its vigorous offspring, the variety Kiftsgate). Many others might claim mention, but we conclude here with *R. mulligani* (Synstylae), shown off to immense advantage in July, as the centerpiece to the White Garden at Sissinghurst in southeast England.

The Impact of Modern Genetics

For plant materials modern genetics embraces nuclear, chloroplast and mito-chondrial genomes. Given the technology available, increasingly we expect two things, firstly an understanding of genome composition in terms of nucleotide se-quences and secondly, that we can intervene directly to alter the genome to suit our purposes. It is with the first of these that this discussion is concerned looking in turn at chloroplast and then nuclear studies. The starting point is the current taxonomy set out in Table 13.1 and the extent, then, to which modern genetics would support it.

cpDNA

Given its high overall nucleotide substitution rate, Matsumoto et al. (1998) utilised the *mat*K gene in some 27 rose accessions sampling the four subgenera and the 10 sections within subgenus Rosa (Eurosa) shown in Table 13.1. In gen-eral, this study supported the conventional arrangement. One anomaly, however, was Pimpinellifoliae, singled out for discussion earlier, and a topic to which we return subsequently.

In another study Takeuchi et al. (2000), compared restriction sites among 32 species. (It should be pointed out that only 11 species were common to this and the previous study and that even where the same species were involved they rep-resent different accessions.) Again, the overall result was to support the familiar classification although *R. hugonis* and *R. spinosissima*, both members of Pim-pinellifoliae occurred in different clusters. The former species was, regrettably from our point of view, only included in the second study.

Nuclear DNA

A third study utilising nuclear ribosomal DNA is that of Wu et al. (2001). Again, different accessions were used and in regard to species there is only par-tial overlap with the previous studies. Here too, however, there was broad sup-port for the arrangement indicated in Table 13.1 except that two species in Pim-pinellifoliae *R. spinosissima* and *R. platycantha*, showed only a relatively weak relationship.

Nuclear genetics has also shed light in a different direction among roses and the degree of genetic variation supposedly available among, for example, modern rose varieties.

At the beginning of this chapter reference was made to various horticultural groups of roses, among them Bourbon, Damask, Gallica and Hybrid Tea. Later, Noisettes, Teas and Hybrid Perpetuals were mentioned. Since rose specialists have continued to perpetuate them these groups are available for genetic analysis. A study by Martin et al. (2001) on these and other groups of roses, using polymorphic DNA allowed an estimate of genetic diversity among them. The most interesting finding was that of genetic diversity among the founding groups narrowing progressively toward the modern hybrid tea rose. The authors write

Our analysis illustrates the limited variability among modern cultivated varieties and the reduced genetic basis of the Hybrid Tea group... and justifies the search for new aesthetical traits, as promoted recently by several breeders notably Austin (1992)

Such a view contrasts with that of Debener et al (1996) who have presented evidence of wide variation, genetically, among modern roses.

Utilising the variation, which does exist, or can be created a new departure in roses has been the assembly of linkage groups utilising, primarily, a molecular approach but incorporating segregation of such characters as single v. double flowers and white v. pink flowers. In such a study Debener and Mattiesch (1999) were able to place 278 of 305 markers from techniques using RAPD and AFLP. Their data is presented for, a diploid rose, as two sets of the seven linkage groups expected.

Rose Evolution

An impression generated by the various studies is how, repeatedly, one genetic technique after another tends to confirm the robust nature of the traditional classification of wild species. The 10 sections of subgenus Rosa genuinely do have some recognisable integrity and, on this basis, the evolution under domestication of the various horticultural types, as understood by Wylie (1954, 1955), is substantially confirmed by the newer approaches.

An interesting problem remains. If the genetic coherence of the various sections is repeatedly underpinned by newer studies, the assumption must be that they are relatively long established and represent substantial departures from some, probably now extinct, wild ancestor. In this sense do we, rather pessimistically, assume that evolution of the various sections is thus beyond our present technologies or do we attempt, by whatever means, to project backwards so as to conjecture some rose ancestral to the present-day groupings?

Rose Conservation

In this, as with other genera, there is a trend to centralise activity, to ensure the material is authenticated, kept in good condition and made available to those having a genuine claim upon it. Gandolin et al (2000) reported on the European network of collections for modern, wild and primitive roses. For China the situation is presently perceived as follows.

China is reckoned to have about 80 roses or 41% of the world total. Table 13.1 indicates the Chinese representation among the various sections. In terms of collecting, the situation is as follows. Zhang (1998) writes:

Hence, the special historical contribution of old China roses was their key germ plasm, through interspecific hybridisation, giving the various cultivar diversities to modern roses.

The matter is seen as, essentially, in the past. The Chinese contribution is already made. The other point of interest is the perceived threat to wild Chinese roses and only two, *R. odorata* and *R. rugosa*, are of special concern, (Fu and Jin 1992). Both are seen as useful germ plasm and the latter has some medicinal significance.

Roses in China Today

A visit to, say, Beijing Flower Market underlines the Chinese interest in horticulture and among the produce are, for example, Hybrid Tea roses bred elsewhere. Given an improving economy, the obvious assumption is that from the point of view of commercial breeding, China will participate increasingly in all aspects of the international trade. In that sense rose breeding will have, so to speak, come home.

Part IV

Conservation and the Environment

14 Chinese Plant Diversity and Its Modern Literature

"The principal object of the Flora of a country is to afford the means of determining (i.e. ascertaining the name of) any plant growing in it. Whether for the purpose of ulterior study or of intellectual exercise... ...the aptness of a botanical description, like the beauty of a work of imagination, will always vary with the style and genius of the author".

George Bentham. *Flora Honkongensis* 1861

From about 1550 to 1750 contact between Chinese and Western plant collectors amounted to cultural coexistence, but with the adoption elsewhere of the Linnean approach, slowly but surely a similar trend developed here. Since, with almost no exceptions, the plant families of China are shared with regions elsewhere, any alternative system could have had, at best, limited appeal. The flora of China has now been subject to Linnean systematics for more than two centuries, the 20th century including an increasing input from Chinese botanists. Many institutions maintain herbaria and any Western taxonomist could feel at home among rows of cupboards, presses and piles of newspapers interleaved with drying specimens.

Botanists' acquaintance with any flora consists initially in describing a succession of new taxa. Eventually it becomes apparent that a published Flora is required as the focus of a common enterprise. With an area the size of China, local Floras have a place and their accretion into a genuinely national one can only be accomplished over a long period. The goals are readily defined. A Flora should be complete for its area for family, genus and species, harmonised with perceptions of these elsewhere and convenient to use. For the United Kingdom with its small size, relatively limited flora and long tradition of interest in plants, this is entirely feasible. For the huge area of China, the process of revision is almost endless, not least because urbanisation and its consequences deplete the flora but also, through introduction, crops and weeds can be added from elsewhere.

Early Western Botanists

It is appropriate to distinguish between those who were primarily collectors and others, who attempted also, some kind of synthesis. Among the latter one of the earliest was Bretschneider (1898), and it is appropriate that not only a genus but one of the very few endemic Chinese plant families is named after him. Following him and incorporating the results of subsequent studies, it is now appropriate in this book to draw together various aspects of Chinese plant life, of which we offer the following appraisal.

Appraising the Chinese Flora

The aim in writing this book has been to emphasise, with various examples, the *issues* or *themes* that run through the study of Chinese plant life. In doing so we have sought to avoid providing an excess of plant names in the interests of clarity. It is now necessary to supply a rather different perspective, which emphasises both how prolific and how diverse is the Chinese flora. Inherent in this, of course, are the themes discussed earlier.

Any overview of the Chinese flora must recognise that China is not only one of the largest countries in the world, but also has, the most diverse flora of any country in the North Temperate Zone. Including the warmer parts of China, in total it has about 30000 species of vascular plants or one eighth of the world total. Even though China is nearly the same size as the continental United States or Europe, it is floristically richer than these two areas combined. While elsewhere we could find a gradient from tropical rainforest to boreal coniferous forest, we need to recognise for China a range of equable climates which have existed for some 15 million years while extinctions due to climatic change have taken place elsewhere. Again, as was indicated earlier, the collision of India with Eurasia brought about the tectonically active, highly dissected and elevated geography of China making the region an important centre not only of survival but, also, one of speciation and evolution. Within this framework it is, then, possible to focus on important details including the following.

Table 14. 1. The numerical significance of major families in the Chinese flora

	The world		China	
	Genera	Species	Genera	Species
Asteraceae (Compositae)	900	25000	200 (17 endemic)	2000
Orchidaceae	735	20000	150 (10 endemic)	1100
Poaceae (Gramineae)	775	10000	230 (15 endemic)	1500

1. The families conspicuously successful around the world contribute significantly to the Chinese flora of which those outstanding are Asteraceae (Compositae), Orchidaceae and Poaceae (Gramineae). (see Table 14.1)
2. Richness in terms of taxonomic diversity can occur in both more primitive and more advanced groups. Among the gymnosperms the fossil record, where available, shows many genera formerly to have been more widely spread even if now they are confined to China. Among the angiosperms for which fossil data is available, it is possible to show recession toward China in some cases

but among the examples given here is seems reasonable to conclude that some have been confined largely to China since their inception. (see Table 14.2.)

Table 14. 2. Primitive and advanced ingredients on the Chinese flora

Gymnosperms	Angiosperms
Cephalotaxaceae (*Cephalotaxus*[a])	Adoxaceae (*Sinadoxa, Tetradoxa*)
Ephedraceae (*Ephedra*[a])	Aristolochiaceae (*Saruma*)
Ginkgoaceae (*Ginkgo*[a])	Bretschneideraceae (*Bretschneidera*[a])
Gnetaceae (*Gnetum*[a])	Cerciodophyllaceae (*Cercidophyllum*[a])
Taxodiaceae (*Metasequoia*)	Davidiaceae (*Davidia*[a])
Pinaceae (*Cathaya*)	Eucommiaceae (*Eucommia*[a])
	Hamamelidaceae (*Fortunearia, Semiliquidambar*)
	Nyssaceae (*Camptotheca*)
	Saruraceae (*Gymnotheca*)
	Tetracentraceae (*Tetracentron*[a])
	Trochodendroaceae (*Trochodendron*[a])

[a]only genus in the family

Apart from the gymnosperm / angiosperm division in Table 14.2 it will be obvious, perhaps, to the reader that both columns have more and less advanced representatives. *Gingko* is primitive relative to, say, *Gnetum*, while *Cercidophyllum* and *Fortunearia* would be considered more primitive than, say, *Saruma* and *Davidia*.

Table 14. 3. Examples of disjunct taxa in the Chinese flora in other locations

Disjunct species	
Liriodendron tulipifera	E. North America, N. Indochina (see Plate 16)
Disjunct genera	
Ancistrocladus	Trop. Africa, Sri Lanka, E. Himal., W. Malaysia
Dillenia	Mascar. S. E. Asia, Indomal, N. Queensl, Fiji
Hamamelis	E. Asia, E. N. Amer.
Illicium	India, S. E. Asia, W. Malaysia, Atlan., N. Amer. Mexico
Lannea	Afri. Indomal
Litsea	N. Asia (Korea, Japan), Austral., Amer.
Nyssa	E. Asia, W. Malaysia, E. N. Amer.
Phoebe	Indomal., trop Amer., N. Indies
Rhodomyrtus	India, Sri Lanka, Thailand, Philip Is., N. Guinea, New Cal. Austr.
Sapindus	Asia, Pacif. (not Austral.), Amer.
Saurauia	Asia, Amer.
Schisandra	Asia, E. N. Amer.
Semecarpus	Indomal, Micron, Solom. Is.

3. To some extent overlapping with the previous consideration, China is rich in endemics, a feature which has been recurrent throughout this book. Ying et al. (1993) record 243 endemic genera encompassing 3116 species. What, though, is of particular interest is the location of concentrations of endemics. These authors recognise three major centres, (A) Western Sichuan – north-western Hubei, (B) Southeastern Yunnan – western Guangxi and (C)Western Sichuan – northwestern Yunnan. Ying et al. (1993) examine the characteristics of each centre in terms of the *kinds* of endemic present from which they infer distinctive evolutionary trends. Their stimulating and thought-provoking discussion is commended to the reader.
4. A feature of the Chinese flora is the presence of a significant number of disjunct taxa – a matter for which examples are set out in Table 14.3.

Table 14.3 shows that different genera can be disjunct between, for example, Asia and N. America, or Asia and tropical Africa. To resolve this, one needs not only an awareness of shifting continents during the period since the rise of the angiosperms but, also, recognition of climatic changes and consequent patterns of migration and extinction over that time.

Given this accumulation of information over a long period, it is necessary to consider the kind of literature it has generated during the modern post-Linnean period and of which the following can only be a small sample.

Primary Literature

Specialist taxonomic and other journals such as appear in the bibliography to this book constitute a primary literature. Unless this is available, the dependent disciplines of anatomy, ecology, genetics, pathology physiology and of course, conservation are impeded from the start.

Secondary Literature

From accumulating published descriptions of plant material and other primary literature it becomes possible, eventually, to generate provincial and national [1]Floras – systematic arrangements of the plants of a given area with a key to their identification and using technically valid plant names. (Primary literature, too, provides a basis for that other kind of secondary literature, textbooks.) An important reference here is that of Keng et al. (1993) *Orders and Families of Seed Plants in China.*

---

[1] The plant content of a given region comprises its flora. The systematic account published in line with taxonomic convention is the Flora.

Table 14.4 summarises the availability of modern Floras of China. A welcome surprise might be how much of it postdates the Cultural Revolution.

Tertiary Literature

From the definitive Floras and associated technical literature, it becomes possible to produce dependent literature more directed toward a popular market. Given China's large and growing tourist industry, such literature, in English and in other languages, is increasingly necessary. Done well, it can not only be attractive, it can alert tourists to the vulnerability of the plant life they have travelled to enjoy. Such literature already exists in profusion and notable examples include those on Chinese alpines (Lang et al. 1997); and Zhu (1999) and on the rare plants of Yunnan (Zhang et al. 1992).

Published Floras Are Provisional

The publication of a regional or national Flora is the outcome of many years' work and involves contributions from perhaps dozens of botanists, whoever might be the final author or editor. Such a publication is a major milestone, but it needs to be recognised that no sooner is it published and in the hands of field botanists than the need for revision has begun. Most of the reasons are obvious and include the following. New records begin to accumulate in that either an unexpected species new to the region or one having a wider range than expected is found. Extinctions, too, can occur or, short of that, recession from the area it once occupied; and, of course, taxonomic opinion changes. One well-known Chinese example concerns *Davidia* Should it be seen as part of Nyssaceae or separated to its own family Davidiaceae? It is not merely a question of naming. If we take Nyssaceae to include *Camptotheca* and *Nyssa* its geographic distribution is E. Asia and eastern N. America. Addition to it of Davidia hardly affects this. *Davidia* separated to Davidiaceae, however, creates a family "entirely confined to China". The first option emphasises relationship and the second distinctiveness – an inescapable theme in taxonomy, and one of which writers of Floras must be aware at species, genus and family level.

As a Flora goes through successive editions or is replaced by a new one altogether, knowledge of that particular region increases but is never complete. Later editions themselves underline the fact of change and, ideally, stimulate their own replacement. The issue is a topical one, since in many parts of the world plant life is under threat through a deteriorating environment. This, in turn, has helped prompt the present-day commitment to conservation, which we now examine in a Chinese context.

Table 14. 4. A Summary of Chinese Floras. (E = English, C = Chinese)

Flora		Dates of Publication	Comment
Flora of China. E.	Vol. 4	1999	Cycadazceae thro. Fagaceae
	Vol. 15	1996	Myrsinaceae thro. Loganiaceae
	Vol. 16	1995	Gentianaceae thro. Boraginaceae
	Vol. 17	1994	Verbenaceae thro. Solanaceae
Flora Reipublicae Popularis Sinicae. C	Vols. 1–65	1959–1997	Concluding volume indexes the set
Provincial Floras. C.			
Anhui	Vols. 1–5	1986–1982	Authorship as 'Edit. Cttee'
Beijing	Vols. 1–2	1984 and 1992	He Shiyuan et al
Fujian	Vols. 1–6	1985–1995	Edit. Cttee.
Guangdong	Vols. 1–3	1987–1995	Authorship as S. China Inst. Bot.
Guangxi	Vol. 1	1991	Authorship as Guangxi Inst. Bot.
Hainan	Vols. 1–4	1964–1977	S. China Inst. Bot.
Hebei	Vols. 1–3	1986–1991	Edit. Cttee
Heilonjiang	Vols. 1,4,5,6&11	1985–1993	Chou Yi-liang et al
Henan	Vols. 1–3	1981–1997	Ding Baozhang et al
Hong Kong	in preparation	–	–
Hubei	1 vol.	1933	
Inner Mongolia	Vols. 2–5	1990–1994	Ma Yu-chun et al
Jiangsu	Vols. 1–2	1977–1982	Jiangsu Inst. Bot.
Jiangxi	Vol. 1	1993	Edit. Cttee. (No angiosperms)
Lianing	Vols. 1–2	1988–1992	Edit. Cttee.
Ningxia			
Qinghai	Vols. 1,3&4	1996–1998	N.W. Plateau Inst. of Biol.
Shandong	Vols. 1–2	1992–1997	Edit. Cttee.
Shanxi	Vol. 1	1992	Edit. Cttee. (No monocotyledons)

Table 14. 4. (cont'd)

Sichuan	Vols. 1–11	1981–1996	Edit. Cttee.
Taijuan	Vols. 1–2	1990–1992	Edit. Cttee.
Taiwan			
Flora of Taiwan (2nd Edn.)	Vol. 1	1984	Edit. Cttee. Flora of Taiwan
"	Vol. 2	1986	Editor in Chief
"	Vol. 3	1993	Hung Tseng Chieng
Xinjiang	Vols. 1–2	1993–1995	Edit. Cttee.
Xizang	Vols. 1–5	1983 –	Wu cheng-yih et al
Yunnan	Vols. 1–8	1977–1998	Kunming Inst. Bot.
Zhejiang	Vols. 0–7	1989–1993	Edit. Cttee.
Floras of Specialised Regions			
Flora in Desertis Reipublicae Popularum Sinarum	Vols. 1–3	1985–1992	Lanzhou Inst. Desert Research
Dulong jiang Region (Yunnan)	1 vol.	1993	Li Heng
Loess - Plateau Sinicae	Vols. 2 & 5	1989–1992	N. W. Inst. Bot.
Mountain Area of Guangdong	1 vol.	1990	Chen Pang yu et al
Nanshua Is. and Neighbouring Is.	Vol. 1	1996	Xing Fuwu et al
Plantarum Herbacearum Chinae Boreali-Orientalis	Vols. 3–7 & 11	1975–1980	Inst. Sylviculturae et Pedologiae
Puyang	1 vol.	undated	Edit. Cttee.
Saihanbaensis	1 vol.	1996	Huang Jinxiang et al
Sinensis in Area Tan Yang	Vols. 1–2	1998–1993	N. W. Inst. of Bot.
Tsinlingensis	Vols. 1–3	1974–1985	N. W. Inst. of Bot.
Mt. Namjagbarwa Region of Xizang	Vol. 1	1992	Ni Zhi-cheng et al
Fossil Floras of China through the Geological Ages. E		1995	Li et al

For further details of publication, the reader is referred, for convenience, to Wanhai Books of Xiangshan, Beijing. Email wanhai@caf.forestry.ac.cn

15 Conservation in Practice

Even before Linnaeus was appointed to his first post, the beginnings of the Industrial Revolution were evident. One might argue, therefore, that the Linnean Project, the cataloguing of the components of the biosphere, began too late, since before we have finished, species have been lost to pollution, urbanisation and the intensification of agriculture. What conservationists have to recognise is that the Industrial Revolution, as an economic engine, has been adopted by almost every country in the world and that it is against such a background we have to sustain, in our own interest, the well–being of the biosphere. We do not have the option of choosing at, say, national level either conservation or industrialisation. We need both.

Conservation in China

This subject is now immense in its scope and complex in terms of how many agencies and their employees are involved. Our aim, therefore, is from the viewpoint of plant science, as elsewhere in this book, to identify themes rather than provide a minutely detailed account.

The first task of the conservationist here, as elsewhere, is essentially protective. Then there is a prospect of rehabilitation of damaged habitats and lastly the possibility of sophisticated management where long–term survival is feasible alongside increasing scientific awareness. All these aspects are evident in modern China and illustrated by reference to the China Plant Red Data Book.

The Chinese Plant Red Data Book

This book. edited by Fu and Jin (1992) and hereafter referred to as CPRDB, represents a major event in Chinese conservation. More than 200 experts from 63 institutions listed 388 taxa as items for conservation. Of these 121 were endangered, that is, in immediate danger of extinction; 110 were rare and, while not at immediate risk, were scattered and recognisably of some concern; 157 were considered vulnerable from natural and/or human causes and might have to be given, at some future date, endangered status. The CPRDB is well illustrated and for each taxon it lists provides information not only as to status but also on morphology, distribution, ecology, biology, protection value, conservation measures, cultivation and references.

To illustrate its use we abstract from it eight families. For the convenience of English speakers, all, save one small exception, Cercidiphyllaceae, are taken from families in already published volumes of the *Flora of China* (Wu and Raven). Both gymnosperms and dicotyledons are included in the selection. Monocotyledons are treated separately in a later section. Table 15.1, which is confined to each family's representation in China, seeks to outline the extent of the threat.

Table 15. 1. Representation in China of eight woody families

	Proportions of threatened taxa	Genera	Species
Gymnosperms			
Cephalotaxaceae	*Cephalotaxus* 3/6	1/1	3/6
Pinaceae	*Pinus* 8/39, *Picea* 6/18, *Larix* 2/11, *Cathaya* 1/1, *Keteleeria* 5/5, *Pseudolarix* 1/1, *Abies* 7/22, *Pseudotsuga* 5/5, *Tsuga* 3/4, *Cedrus* 0/2	9/10	38/108
Taxodiaceae	*Cunninghamia* 1/2, *Glyptostrobus* 1/1, *Taiwania* 1-2/1-2, *Cryptomeria* 0/1, *Taxodium* 0/2, *Sequoiadendron* 0/1, *Sequoia* 0/1, *Metasequoia* 0/1	4/9	4/11
Angiosperms			
Cercidiphyllaceae	*Cercidiphyllum* 1/1	1/1	1/1
Juglandaceae	*Platycarya* 0/1, *Engelhardtia* 0/4, *Cyclocarya* 0/1, *Pterocarya* 0/5, *Juglans* 2/3, *Annamocarya* 1/1, *Carya* 0/5	2/7	3/20
Oleaceae	*Fontanesia* 0/1, *Fraxinus* 1/32, *Forsythia* 0/6, *Syringa* 1/16, *Osmanthus* 0/22, *Chionanthus* 0/7, *Olea* 0/13 *Myxopyrum* 0/2, *Ligustrum* 0/27, *Jasminum* 0/43	2/10	2/159
Sapotaceae	*Madhuca* 2/2, *Manilkara* 0/1, *Palaquium* 0/1, *Eberhardtia* 0/2, *Chrysophyllum* 0/1, *Diploknema* 0/2, *Xantolis* 0/4, *Pouteria* 0/2, *Planchonella* 0/2, *Sinosideroxylon* 0/3, *Sarcosperma* 0/4	1/11	2/24
Styraceae	*Styrax* 0/31, *Bruinsmia* 0/1, *Alniphyllum* 0/3, *Huodendron* 0/3, *Halesia* 1/1, *Melliodendron* 0/1, *Pterostyrax* 1/2, *Sinojackia* 2/5, *Rehderodendron* 1/5, *Parastyrax* 0/2	4/10	5/54

From Table 15.1 it is evident that:
1. For a family, threat can involve one to several genera.
2. For a given genus, from one to several species can be threatened.

3. Taken overall, any proportion of a family might be threatened. This is, relatively, large in Pinaceae, almost negligible in Oleaceae and, curiously, in Cercidiphyllaceae, through one species, the entire family.

Not evident from Table 15.1 is the emphasis for conservation upon the endemic part of the flora.

From Table 15.1 it is possible to abstract manageable groups of either genera
or species to illustrate the various concerns of the conservationist for these families.

Cephalotaxaceae

Of its six species in China, three are threatened. Two particularly, *C. mannii* and
C. oliveri, are heavily exploited medicinally. Since this genus is dioecious, the
more scattered a population becomes the greater is the problem of reproduction.

Pinaceae

1. *Abies*. Of the various species in China seven are under some threat. Among
   them are the problems of rarity, poor regeneration and loss of habitat. Beyond
   this, a feature of especial interest is their distribution. *A. beshanzuensis* is confined to five known specimens in east China, *A. fanjingshanensis* in a restricted endemic in Guizhou, *A. sibirica* is known only in the north west on
   the Altai mountains and *A. yuenbaoshanensis* apparently consists of only 100
   mostly old individuals confined to Guangxi for example. Such distributions
   are reckoned to be significant in terms of probing climatic change and are
   therefore of considerable importance.
2. *Keteleeria, Larix, Picea* (Plate 7), *Pinus, Pseudolarix* (Plate 16), *Pseudotsuga*
   and *Tsuga*. The problems and interests indicated for *Abies* are essentially similar for these genera. A theme throughout is that, given adequate supplies, virtually all of them could yield useful timber.
3. *Cathaya*. This was singled out for substantial comment in an earlier chapter.
   The interest here is that it is a monospecific genus whose pollen has been
   found in the Tertiary sediments of Eurasia. Its morphology makes it of great
   interest in the phylogeny of the Pinaceae.

Taxodiaceae

Of the four genera considered threatened, *Glyptostrobus* and *Metasequoia* are
monospecific, *Cunninghamia* (Plate 16), thought so in China until the recognition of *C. unicaniculata* and *Taiwania* as either monospecific or consisting of two
species, depending on various authorities. *C. unicaniculata* and *Taiwania* pro-

vide timber but the others, because of their rarity, are at present, chiefly of scientific interest. They provide, through their fossil record, evidence of a far wider distribution and, a familiar theme not merely among conifers, of survival in China following elimination elsewhere.

Cercidiphyllaceae

Cercidiphyllum japonicum is a well known ornamental cultivated in various parts of the world and on that account hardly threatened. What is of more interest is its natural distribution in China and Japan where it provides evidence of a link between the floras of the two countries reaching back to the Tertiary.

Juglandaceae

1. *Annomocarya sinensis* is a rarity found in southwest China of some interest in understanding the phylogeny of its family. Provided supplies are considerably increased, it could be significant as a high–quality timber and a source of an industrially useful oil.
2. *Juglans mandshurica* and *J. regia.* The former is a timber and a source of grafting stock for commercial walnuts. The latter, existing in perhaps a 1000 or so wild trees, is confined to Xinjiang and is seen as a precious Tertiary relict important in the study of palaeogeography. Cultivars, of course, are widely grown.

Oleaceae

1. *Fraxinus mandshurica.* Scattered remnants exist in northeast China and the primary interest is to retrieve it and multiply stocks so as to make its high–quality timber available.
2. *Syringa pinnatifolia.* The genus is familiar to Western horticulturalists, but this species is of little such interest. Primarily, the concern is for an endangered endemic thought significant in the distribution and phylogeny of the genus in China. The species' medicinal reputation has reduced it to the verge of extinction.

Sapotaceae

1. *Madhuca hainanensis* and *M. pesquieri.* The two species are distributed with the former in Hainan and the latter in southern China. The reasonable assumption is that they have a close affinity, influenced, perhaps, by whatever isolation through time Hainan has afforded although, as discussed earlier, it

was for a time joined to the mainland. Both are high–quality timber trees and the latter supplies, from its seeds, an edible oil.

Styracaeae

1. *Halesia macgregorii*. The genus is disjunct, occurring in southeast China and North America, and is therefore a phytogeographic puzzle. In China it is rare with poor regeneration. The American relatives notably *H. monticola* var. *vestita* are esteemed by Western gardeners. This Chinese representative is seen as significant both for timber and as an ornamental.
2. *Pterostyrax psilophylla*. The genus reaches from Burma to Japan and, although vulnerable, is considered valuable as a garden ornamental. The species is already available in Western gardens.
3. *Rehderodendron macrocarpum*. In some respects, though not in its disjunct distribution, its situation recalls *Halesia*. Though rare in China in the wild, it is readily available as an ornamental in the West.
4. *Sinojackia dolichocarpa* and *S. xylocarpa*. The genus is endemic to China, the former species only relatively recently discovered. Its species are reduced to small populations with poor genetic prospects through habitat destruction. Another species, *S. rehderiana*, is known in Western gardens.

This exploration of eight families reiterates the themes of a flora under threat, habitat destruction, overexploitation and of course, if action is taken in time, all manner of commercial advantages. There are also the recurrent interests of evolution palaeoclimatology and plant geography. A more detailed survey of the CPRDB would only reveal how widespread these problems are. We now consider, on a different basis, the monocotyledons.

Monocotyledons

In the CPRDB six families are included in this group – Gramineae (Poaceae), Hydrocharitaceae, Liliaceae, Musaceae, Orchidaceae and Palmae (Arecaceae). Apart from some rices in Gramineae, the other genera there are bamboos and, with the Palmae comprise a more woody group whose situation is not dissimilar to tree taxa in the previous section, and are thus excluded from the present discussion. Of the remaining four families, sexual or vegetative reproduction is mostly conveniently available and propagation under artificial conditions would usually be straightforward.

As to the nature of the threats, in Hydrocharitaceae, a water plant, *Ottelia accuminata* is subject to pollution. In Liliaceae *Dracaena*, *Fritillaria* and *Trillium* are overcollected for medicinal purposes. The genus *Orchidantha* is of uncertain affinity attributed in CPRDB to Musaceae and by others to the Lowiaceae. *O.*

chinensis was known at two locations only in Guangxi in 1992. It poses the familiar problem of a rare plant being sought for medicinal use. Among Orchidaceae, the problems are those of habitat destruction, overexploitation medicinally and, also, removal by acquisitive orchid enthusiasts.

A Wider Perspective

The rich plant life of China is of immense horticultural interest in other countries. Especially sought–after genera are *Aconitum, Androsace, Bergenia, Camellia, Cardiocrinum, Clematis, Daphne, Dianthus, Epimedium, Gentiana, Hosta, Meconopsis, Nomocharis, Paeonia, Primula, Rhododendron, Rosa, Roscoea* and, of course, many others and here, too, it is important to avoid over-collection and introduce protective measures. Properly regulated, a legitimate trade offers China a long-term commercial opportunity. For a comprehensive list of suitable species see Feng (1999).

Organisation

Hitherto, in this chapter, the focus has been on the plants themselves and the scientific and associated issues they prompt. It will, however, be obvious that there needs to be administrative and technical support and, underlying this, an appropriate legal framework at local, provincial and national level. For a book of this nature primarily of botanical interest, it is inappropriate to consider this in great detail, but a summary is required and is as follows.

China has in recent years seen a steady improvement in policy toward the environment and in the laws and regulations to control utilisation of natural products and promote their conservation. This country has also become increasingly visible in international conservation efforts. A convenient source of detail in these matters is Zhang (1998). This editor, for example, includes legislative works and international treaties involving China (notably concerning the Antarctic, trade in endangered species – CITES, conventions on biodiversity, wetlands and desertification among others). Included, too, is a classified catalogue of nature reserves, which are then individually listed, totalling 799, and by 1995 covering a total of 71906710 ha (by 1997 there were 926 reserves covering 77 m. ha or 7.64% of the total national land area. It is intended eventually that this should rise to 10%.) Nearly 60 zoos are listed. For the readers of the present book, of particular interest are details of some 57 botanical gardens distributed across China, and these are individually listed in Appendix 2.

It is necessary to recognise two things. The first is the sheer size and complexity of the Chinese biota within a vast and diverse landscape. The second is to allow that conservation is only one of modern China's preoccupations. An anecdote from one of us (G.P.C.) is appropriate regarding botanical skills in unexpected places.

"Diverting from the Silk Road, we drove up rough tracks to eventually alpine pasture grazed by yaks. The manager of the site emerged from a bleak concrete block office, seemingly in the middle of nowhere, and in response to our enquiry identified for us in the grass sward *Elymus nutans, Gentiana macrophylla, G. dahurica, Anaphalis lactea, Lomatogonium carinthiacum* and *Stellare chamaejasme*, remarking that this last one was very poisonous. We had met him only by chance and, given his likely priorities, I cannot imagine the floristic content of alpine pasture was high among them".

Biodiversity

Each of the various topics raised throughout the present book can be seen as a tributary issuing into the larger river of biodiversity. A botanist cannot be unaware of animals. A conservationist needs to recognise the interests of both hunters and birdwatchers. Who could seriously propose an animal species be conserved in the wild without ensuring the existence and maintenance of its habitat? Living matter is essentially *interactive* and we isolate some aspect of it only for the convenience of study. Sooner or later what we have learned has to be integrated into a wider understanding.

A Further Landmark

The CPRDE (1992), a major event in conservation, was followed by another in 1998 the publication of *China's Biodiversity: a Country Study* (edited by Zhang), and from which the organisational data referred to here was taken. Here, the various agencies of the State have demonstrated practical interaction with scientists working in the field on the dynamics of living matter. The book is immensely informative about the practicalities of sustaining biodiversity in China and provides an essential background for anyone wishing to study plants there. To illustrate something of its interest and usefulness, the following examples are selected here. Essentially, the threats perceived in 1992 remain and there is, detectably, an increased sense of concern as, to the earlier problems, new ones have been added. So severe is the threat to *Cephalotaxus* perceived to be, that for *C. hainanensis* protection has been strengthened – primarily because it has been found to be a source of anti-carcinogens. Again, four new species of *Cycas, C. panzhihuaensis, C. guizhouensis, C. multipinnata* and *C. micholitzii*, were destroyed immediately after they were announced as new species to the public or their distribution sites were discovered. Tourism is both an advantage economically and, for the conservationist, can be a disadvantage. The orchid *Gymnadenia conopsea*, 2100 m above sea level in a reserve in Hebei, is on the verge of extinction because of collection by tourists. Doubtless, this problem will only increase. Among the other new concerns is the release of genetically modified organisms.

What is evident from the report is a real concern for the biodiversity of China toward which is directed the skill and commitment of so many experts. There are, of course, all manner of difficulties to be faced. Consider the following poignant comment from Zhang (1998):

"Vast areas of land with its forests and grasslands have been reclaimed, resulting in the destruction of wild medicinal plant resources. For example, Jianqiao in Hangzhou, Zhejiang Province, used to be the cultivation base for medicinal plants such as *Rehmannia glutinosa* var. *lutea* and *Ophiopogon japonicus*. It was, however, ruined by the pace of industrialisation and the germ plasm resources have disappeared".

It encapsulates the conservationists' difficulty – how to sustain biodiversity alongside industrialisation (and eventually tourism). It emphasises, too, the need for both proper control measures and lively, penetrating programmes of public education. It is not, however, uniformly depressing, and the remarkable successes now of China's panda breeding programme underline what can be achieved when resources and commitment become available.

Collection: an Appraisal

It is necessary to be realistic about plant collection. There is, for example, an indigenous tradition stretching back over thousands of years of collecting plants for medicinal purposes and as population increases and vegetation cover decreases, it is self-evident that cultivation must eventually replace collection from the wild. To achieve this requires recognition of the fact among conservationists, the provision of facilities and public education.

Added to this tradition was the arrival of the Western plant collectors referred to in Chapter 4. Some of them came under commercial sponsorship but their interests were scientific, too, and much of their work is now part of both technical and popular literature. Even the most ardent collectors among them made a negligible impact on the vegetation of China. What they collected has passed into the care of botanic gardens and nurserymen. The dawn redwood, for example, is, through vegetative propagation, widely available in the West and poses no threat whatever to stocks of the plant in China. The same is true of hundreds of other species derived originally from China.

In complete contrast to this is the arrival in China of increasing numbers of tourists, some of them keen to see the plant life that awaits them and, is some cases, either through greed or ignorance, happy to smuggle out rare or striking specimens. Hong Kong, for example, in the public interest, sends contraband orchids to the Kadoorie Foundation for safekeeping (Plate 16). The overall problem is to display China's remarkable plants while ensuring that they are properly safeguarded.

Finally, there is collection by scientists operating under proper supervision and in accordance with international conventions. If, for example, a rare plant is discovered and elsewhere is a specialist tissue culture laboratory with the particular skills to propagate it, it makes sense to share it with, perhaps, the prospect eventually of reintroducing it to its original habitat. Such collection is, quite positively, in aid of the maintenance of the flora's diversity and interest.

Concluding Retrospective

In a book of this length, it is not possible to say everything, but rather the aim has been to raise interest and awareness. How then are we to conclude matters? The message surely is twofold. The first part is that China contains a flora of extraordinary diversity and interest able to challenge botanists, delight horticulturalists and through their combined effort, enrich agriculture and the world's gardens. The second part is more sombre. It is no longer possible for any of us to take the existence of such a flora for granted, and the closing sentence of this book must leave the reader in no doubt about the *vulnerability* of China's plant life.

Appendix 1

A Summary of Famine Foods Listed in the *Chiu Huang Pen Tsao* (1406 A.D.)

Notes:
1. The list on which this appendix is compiled is based on Read (1946).
2. Read's original binomials are retained to facilitate cross reference to Vavilov (1920) and because they provide the starting point for comparison with modern taxonomic revisions.
3. Obvious spelling mistakes have been corrected.
4. To highlight the diversity Read's original alphabetical list has been rearranged reflecting familial relationships.
5. Where V precedes a binomial this indicates that it was mentioned by Vavilov (1920).
6. From Read's list of what is used it is not invariably clear what is intended and in these cases there is a ? in place.
7. Read claimed to have identified 358 Chinese foods of the 414 listed in the original work.
8. To indicate the food usage binomials are followed by abbreviations thus:

Bb-Bulb	R - Root
Fl - Flower	Rh - Rhizome
Fr - Fruit	Se - Seed
L - Leaf	Sh - Shoot
Le - Legume	St - Stem
IB - Inner Bark	T - Tuber

In the opinion of the present authors, given the renewed interest in early Chinese agriculture and the very considerable taxonomic revision that has occurred with the Chinese flora since 1946, it would be appropriate for there to be a substantial reworking of the *Chiu Huang Pen Tsao* as regards our present–day understanding of it, not simply out of concern for botanical up–dating but because the document is part of agricultural and sociological history. *Ti shen*, Read calls bluebell and attributes to it two quite unrelated genera *Adenophora* (Campanulaceae) and *Anemarrhena* (Liliaceae). At least as surprising is *Ch'iao Mai* or Brome grass to which Read attributes not only *Bromus* (Poaceae) but also *Fagopyrum* (Polygonaceae)! *Diospyros kaki*, for example, is now recognised to exist in three varieties. The list could be multiplied but the primary object here is to provoke the reader's curiosity and interest.

LICHEN			Dioscoreaceae		
Cladonia V. Doubtful (Read).			*Dioscorea*		
FERN				*japonica*	T, Se
	Polypodiaceae		Liliaceae		
	Drymoglossum			*Aletris*	
		carnosum	Sh, L		*japonica*	L
		subcordatum	Sh, L		*spicata*	L
	Cynostemma			*Allium*	
		pedata	L		*takeri*	L, Bb, Fr
CONIFER			V	*fistulosum*	Sh, St
	Cupressaceae			*nipponicum*	St, R
	Thuja			*odorum*	St, L
		orientalis	Se		*victorialis*	St, L
				sp	
MONOCOTYLEDONS				*Anemarrhena*	
Alismataceae				*asphodeloides*	R
V	*Alisma*		V	*Asparagus*	
		plantago	L		*lucidus*	T
V	*Sagittaria*			*Hemerocallis*	
		sagittifolia	Sh		*flava*	Fl, L, Sh, R
Amaryllidaceae				*Lilium*	
	Lycoris			*brownii*	Bb
		aurea	R		*Liriope*	
		radiata	R		*spicata*	T
Araceae				*Ophiopogon*	
	Acorus			*japonicus*	T
		calamus	Rh		*Polygonatum*	
	Colocasia			*falcatum*	Rh
		antiquorum	R		*giganteum*	Sh, R
Butomaceae				*multiflorum*	Rh
	Butomus			*officinale*	Rh
		umbellatus	R		*Scilla*	
Commelinaceae				*chinensis*	Bb
	Commelina			[1]*japonica*	Bb
		communis	Sh, L	Poaceae		
Cyperaceae				*Brachypodium*	
	Eleocharis				
		plantaginea	St			
	Fimbristylis				
		sub -bispicata	Sh, R			
	Mariscus				
		sieberianus	R, Se			

[1] Read remarks that it must be very thoroughly soaked and boiled. He adds that eating this bulb produces gas and rumbling in the belly.

	japonicum	Se	Actinidiaceae	
Bromus			*Actinidia*	
	agrestis	Se	*chinensis*	L, Fr
	japonicus		Amaranthaceae	
Coix			*Achryanthes*	
	lachryma - jobi	Se	*aspera*	Sh, L
Eriochloa			*bidentata*	Sh, L
	villosa	Se	*Amaranthus*	
Imperata			*blitum*	St, L
	arundinacea	Sh	*tricolor*	St, L
Panicum			*Celosia*	
	crus-galli	Se	*argentea*	Sh, L
	frumentaceum	Se	Anacardiaceae	
Phragmites			*Rhus*	
	communis	Sh	*cotinus*	Sh
V	*Phyllostachys*		Araliaceae	
	bambusoides	Sh	*Acanthopanax*	
Setaria			*recurvifolium*	Sh, L
	glauca	Se	*ricifolium*	Sh, L
Zizania			V *Aralia*	
	aquatica	Se, Sh	*cordata*	L
Potamogetonaceae			Aristolochiaceae	
Potomogeton			*Aristolochia*	
	crispus	L, R	*debilis*	L, Fr, St, R
Smilacaceae			Asclepiadaceae	
V *Smilax*			*Cynanchum*	
	china	L	*atratum*	L, Fr
	herbacea	L	[2]*caudatum*	L, R
	seiboldii	Sh, L	*sibiricum*	Se
	trinervula	Fr	*Marsdenia*	
Typhaceae			*tomentosa*	Sh, L
Typha			*Metaplexis*	
	japonica	Sh, R	*stauntonii*	L
	latifolia	Sh, R	Balsaminaceae	
			Impatiens	
DICOTYLEDONS			*balsamina*	Sh, L
Acanthaceae			Berberidaceae	
Hygrophila			*Epimedium*	St, L
	lancea	Sh, L	Bignoniaceae	
	quadrivalis	Sh, L	*Catalpa*	
Rostellularia			*kaempferi*	Fl
	procumbens	L	*Incarvillea*	
Strobilanthes			*delavayi*	L, Sh, St
	oliganthus	Sh, L	*sinensis*	L, Sh, St
Aceraceae			Boraginaceae	
Acer				
	palmatum	L		
	pictum	L		

[2] Read states that the root contains cynano-
chotoxin a chemical causing paralysis.

Eritrichium	
pedunculare	Sh, L
Calycanthaceae	
Chimonanthus	
fragrans	Fl
Campanulaceae	
Adenophora	
polymorpha	Bb, L, Sh, R
remotifolia	R
stricta	L, Sh, R
verticillata	Bb
Campanula	
punctata	L
Lobelia	
sessilifolia	L
Platycodon	
grandiflorum	L
Wahlenbergia	
gracilis	R
Cannabinaceae	
V [3]*Cannabis*	
sativa	L, Se
Humulus	
japonicus	L, Sh
Caprifoliaceae	
Lonicera	
gracilipes	Fr
[4]*japonica*	
[5]*morrowii*	Fr
Viburnum	
dilatum	Fr
japonicum	L
sempervirens	L
Caryophyllaceae	
Cerastium	
triviale	L, Sh
Cucubalis	
baccifer	L
Dianthus	
superbus	Sh, L
Saponaria	

[3] Read asserts that the tall Chinese variety of hemp lacks the narcotic effect of shorter Indian varieties.
[4] supposedly emetic and cathartic.
[5] The dry stem yields pharmacologically active substances.

vaccaria	L, Se
Silene	
aprica	Sh, L
Stellaria	
aquatica	Sh, L
Chenopodiaceae	
Agriophyllum	
arenarium	Sh, L
Achroglochin	
persicarioides	L
Beta	
vulgaris	St, L
Beta	
vulgaris var. *cicla*	
Chenopodium	
Kochia	
album	Sh, L, Se
scoparia	Sh, L
Suaeda	
glauca	Sh, L
Clethraceae	
Clethra	
barbinerva	L
Compositae	
Adenocaulon	
bicolor	L
V *Arctium*	
lappa	L, R
Artimesia	
dracunculus	L
keiskiana	Sh, L
lavendulaefolia	L
stelleriana	Sh, L
vulgaris	L
sp.	
Aster	
indicus	L
trinervius adustus	Sh, L
trifolium	Sh, L
Asteromoea	
cantonensis	L
Atractilis	
ovata	R
Cacalia	
aconitifolia	L
Calendula	
officinalis	Sh, L
Carpesium	

abrotanoides	Sh, L	*farfara*	L
cernum	L	*Xanthium*	
Carthamus		*strumarium*	Sh, L, Fr
tinctorius	L	Convolvulaceae	
V *Chrysanthemum*		*Calonyction*	
coronarium	St, L	*speciosum*	L, Se
segetum	Sh, L	*Calystegia*	
sinensis	Fl, L	*hederacea*	R
Cnicus		*japonica*	L, Sh, R
japonicus	Sh, L	*sepium*	R
spicatus	Sh, L	Cornaceae	
Crepis		*Cornus*	
japonica	St, L	*macrophyllus (ident?)*	
Echinops		*officinalis*	Fr
dahuricus	L	Crassulaceae	
Eclipta		*Sedum*	
alba	Sh, L	*japonicum*	L
Elephantopus		*kamtschaticum*	Sh, L
scater	L	*luneare*	L
Gnaphalium		Cruciferae	
japonicum	Sh, L	*Arabis*	
sp?	Sh, L	*glabra*	St, L
Heteropappus		*perfoliata*	St, L
hispidus	L	*Brassica*	
Inula		*campestris*	St, L
britannica	L, Fl	*Capsella*	
chinensis	L	*bursa-pastoris*	L, Se
Lactuca		*Eutrema*	
brevirostris	St, L	*hederaefolia*	
debilis	Sh, L	*wasabi*	Sh, L
denticulata	St L	*Isatis*	
denticulata var *sororia*	St, L	*tinctoria*	L
lacinata	St, L	*Lepidium*	
Saussurea		*sativum*	Sh, L
affinis	Sh, L	*virginicum*	Sh, L
Scorzonera		*Nasturtium*	
albicaulis	Sh, L	*globosum*	Sh, L
hispanica	Sh, L	*indicum*	Sh, L, R
Senecio		*montanum*	St, L
palmatus	L	*officinale*	Sh, L
Sigesbechia		*palustre*	Sh, L
orientalis	Sh, L	*Sisymbrium*	
Sonchus		*sophia*	Sh, L
oleraceus	St, L	*Thlaspi*	
Tanacetum?		*arvense*	L
Taraxacum		Cucurbitaceae	
officinale	St, L	*Cucurbita*	
Tussilago		*pepo*	R

	Gymnostemma	
	pedatum	?
V	*Luffa*	
	cylindrica	Fr
	Melothria	
	japonica	Fr
	Momordica	
	charantia	Fr
	Trichosanthes	
	japonica	Fr, R
	kirilowii	Fr, R
Dipsacaceae		
	Scabiosa	
	japonica	R
Ebenaceae		
V	*Diospyros*	
	kaki	Fr
V	*lotus*	Fr
Eleagnaceae		
	Eleagnus	
	longipes	Fr
Euphorbiaceae		
	Euphorbia	
	helioscopia	Sh, L
	humifusa	Sh, L
	Phyllanthus	
	urinaria	L
Eupteleaceae		
	Euptelea	
	franchetii	L
	polyandra	L
Euryalaceae		
	Euryale	
	ferox	P, Se
Fagaceae		
	Quercus	
	bungeana	Se
	glauca	L
	[6]*tungeana*	Se
	sp.	
Fumariaceae		
	Corydalis	
	incisa	

Gentianaceae		
	Gentiana	
	scabra	L, R
Geraniaceae		
	Geranium	
	nepalense	L
Helwingiaceae		
	Helwingia	
	rusciflora	Sh, L
Juglandaceae		
	Juglans	
	regia	Se
	Platycarya	
	strobilacea	Sh, L
Labiatae		
	Calamintha	
	chinensis	L
	Leonurus	
	sibiricus	Sh, L
	Lycopus	
	europaeus	R
	lucidus	R
	Mentha	
	arvensis	St, L
	Nepeta	
	japonica	St, L
	tenuifolia	St, L
	Ocimum	
	basilicum	St, L
	Perilla	
	nankinensis	L, Se
V	*ocimoides*	St, L, Se
	Plectranthus	
	longitubus	L
	Prunella	
	vulgaris	L
	Salvia	
	japonica	L
V	*Stachys*	
	sieboldii	L
Lardizabalaceae		
	Stauntonia	
	hexaphylla	Fr
Leguminosae		
[7]	*Albizia*	
	julibrussin	L

[6] Read comments that the bitter principle removed by repeated soaking and them steaming is a cause of diarrhoea. What remains is highly nutritious, allegedly.

[7] The stem bark is a source of saponins

	Apios	
	fortunei	T
V	*Astragalus*	
	henryi	Sh, L
	hoantehy	Sh, L
	sinicus	Sh, L, Se
	Canavalia	
	ensiformis	Sh, L, Fr, Se
	gladiata	Sh, L, Fr, Se
	Caragana	
	chamlagu	Fl
	Cassia	
	mimosoides	Le, Se
	sophora	Fl, L, Sh
	Dalbergia	
	hupeana	Sh, L
	Desmodium	
	japonicum	Fr
	podocarpum	Fr
	Dolichos	
	lablab	Sh, L, Fr, Se
	Dumasia	
	truncata	L, Fr, Se
	Gleditsia	
	japonica	Sh, Se
	sinensis	Sh, Se
V	*Glycine*	
	hispidia	Sh, L, Fr, Se
	soja	Fr, Se
	ussuriensis	Se
	Indigofera	
	decora	Se
	pseudo tinctoria	Fl
	Lathyrus	
	maritimus	Se
	palustris	Le
	Lespedeza	
	bicolor	L, Se
	juncea	Sh, L
	macrocarpa	L, Se
	striata	Se
	Medicago	
	denticulata	L, St
	Phaseolus	
	mungo	L, Le
	Pueraria	
	hirsuta	Fl, R

	[8]*Sophora*	
	japonica	L, Sh, Fl
	Swainsonia	
	salsula	L, Se
	Trigonella	
	caerulea	Sh, L
	Vicia	
	faba	Fr, Se
	unijuga	L
V	*Vigna*	
	sinensis	Sh, L, Fr, Se
	var	?
	Wisteria	
	chinensis	Fl
Lentibulariaceae		
	Utricularia	
	vulgaris	Sh, L, Fr, Se
Lythraceae		
	Lythrum	
	anceps	St, L
	salicaria	St, L
	Rotala	
	indica	Sh
Malvaceae		
V	*Abutilon*	
	avicennae	Se
	Hibiscus	
	mutabilis	L
	syriacus	L
	trionum	Sh, L
	Malva	
	verticillata	St, L
Meliaceae		
	Cedrela	
	sinensis	Sh
Menyanthaceae		
	Limnanthemum	
	nymphoides	Sh
Moraceae		
V	*Broussonetia*	
	papyrifera	Fl, Fr
	Cudrania	

[8] Although the *Chiu Huang Pen Tsao* considers the shoots, leaves and flowers edible, it is these last when dried which yield rutin and quercetin, both now known to be mutagenic.

	triloba	L, Fr
	Ficus	
	carica	Fr
V	*Morus*	
	alba	L, Fl, Fr
Nelumbonaceae		
V	*Nelumbo*	
	nucifera	R, Se
Oleaceae		
	Forsythia	
	suspensa	L, Fr
	Ligustrum	
	japonicum	Sh, L
	lucidum	Sh, L
Onagraceae		
	Epilobium	
	hirsutum	St, L
	macranthum	L
	pyrricholofolum	L
Oxalidaceae		
	Oxalis	
	corniculata	Sh, L
Papaveraceae		
	Chelidonium	
	majus	L
V	[9]*Papaver*	
	somniferum	L, Se
Pedaliaceae		
V	*Sesamum*	
	indicum	Sh, L, Se
Penthoraceae		
	Penthorum	
	chinense	Sh, L
	sedoides	
Phytolaccaceae		
	[10]*Phytolacca*	
	acinosa	R

Plantaginaceae		
	Plantago	
	major	Sh, L
Polygalaceae		
	Polygala	
	japonica	L, Sh, R
	[11]*tenuifolia*	L, Sh, R
Polygonaceae		
V	*Fagopyrum*	
	esculentum	Sh, L, Se
	Polygonum	
	aviculare	Sh, L
	cuspidatum	Sh, L
	hydropiper	St, L
	multiflorum	Fl, R
	orientale	Sh, L
	persicaria	St, L
	Rumex crispus (Read's missing entry 7.21)	
Portulacaceae		
	Portulaca	
	oleracea	St, L
Primulaceae		
	Lysimachia	
	candida	Sh, L
	clethroides	Sh, L
	fortunei	Sh, L
Punicaceae		
	Punica	
	granatum	L, Fr
Ranunculaceae		
	Aquilegia	
	flabellata	L
	[12]*Clematis*	
	angustifolia	L
	chinensis	L, R
	paniculata	Fl, L
	Ranunculus	

[9] The seeds and leaves of the poppy contain none of the narcotic prinicples of opium. (Read)

[10] Leaves used as a vegetable in the Himalayas and Japan. The root contains the very toxic substance phytolaccotoxin. To render safely edible required considerable preparation involving slicing, boiling and soaking. (Read)

[11] The core of the root is deleterious and should be removed. (Read, significantly, comments that the root has chemicals shown to be pharmacologically active.)

[12] Read comments that the plant contains anemonine and that the repeated boiling and soaking required in successive changes of water might leave little nutriment.

	japonicus	L
	pennsylvanicus	L
Rhamnaceae		
	Hovenia	
	dulcis	Fr
	Rhamnus	
	virgatus	L
V	Zizyphus	
	vulgaris	L, Fr
Rosaceae		
	Agrimonia	
	eupatoria	Se
	Crataegus	
	cuneata	Fr
	pinnatifida	Fr
	Cydonia	
	sinensis	Fr
	Fragaria	
	indica	Fr
	Photinia	
	villosa	L
	Potentilla	
	chinensis	Sh, L
	discolor	R
V	Prunus	
	armeniaca	
	communis	Fr
	japonicus	Fr
V	mume	Fr
V	persica	L, Fr
V	pseudocerasus	Fr
V	tomentosa	Fr
	undulata	Fr
	Pyrus	
	betulaefolia	L, Fl, Fr
	serotina	L, Fr
	var culture	
	sinensis	L, Fr
	Rosa	
	indica	Sh, L
	Rubus	
	thunbergii	Fr
	Sanguisorba	
	minor	Sh, L
	officinalis	L

	[13]Sorbus	
	aucuparia	Sl, L
Rubiaceae		
	Galium	
	verum	Sh, L, Se
V	Rubia	
	cordifolia	Fr, R
Rutaceae		
	Zanthoxylum	
	piperitum	L
Salicaceae		
	Populus	
	alba	L
	Salix	
	gracilistyla	L
Sapindaceae		
	Koelreuteria	
	paniculata	Sh, L
	Xanthoceras	
	sorbifolia	L, Fl, Se
Scrophulariaceae		
	[14]Rehmannia	
	glutinosa	L, R
	Veronica	
	agrostis	Sh, L
	anagallis	Sh, L
	longifolia	Sh, L
	spuria	Sh, L
Simaroubaceae		
	Picrasma	
	quassioides	Sh, L
Solanaceae		
	Lycium	
	chinense	Fr
	Physalis	
	alkekengi	L, Fr
	Solanum	
	nigrum	L, Fr
	septumlobum	L

[13] Read notes that the leaves contain amygdalin a cyano-genetic glucoside.

[14] The root described here as a famine food is also a traditional Chinese medicinal material. Read comments (not perhaps surprisingly) that the root is steamed and sundried nine times.

Sparganiaceae		
Sparganium		
longifolium	Se	
Staphyleaceae		
Staphylea		
bumalda	L	
Styracaeae		
Halesia		
corymbosa	L	
Theaceae		
Thea		
sinensis	L	
Tiliaceae		
Tilia		
argentea	L	
oliveri	L	
Trapaceae		
Trapa		
natans	Se	
Ulmaceae		
Celtis		
sinensis	L, Fr	
Ulmus		
campestris	L, Se, IB	
Umbelliferae		
Apium		
graveolens	Sh, L	
Bupleurum		
falcatum	Sh, L, R	
Cnidium		
monneiri	Sh, L	
Conioselinum		
univittatum	L	
Cryptotaenia		
japonica	L	
Foeniculum		
vulgare	Sh, L, Se	
Nothosmyrnium		
japonicum	Sh, L	
Oenanthe		
stolonifera	L	

Osmorhiza		
aristata	R	
japonica	R	
Peucedanum		
decursivum	L, R	
Sanicula		
europaea	Sh, L	
sinensis	Sh, L	
Seseli		
libanotis	L	
Siler		
divaricatum	Sh, L	
Urticaceae		
V *Boehmeria*		
nivea	R	
Valerianaceae		
Patrinia		
palmata	L	
Verbenaceae		
Callicarpa		
mollis	Fr	
Vitex		
incisa	Se	
negundo	Se	
Violaceae		
Viola		
verecunda	Sh, L	
Vitidaceae		
Ampelopsis		
heterophylla	L	
Vitis		
labrusca	Fr	
thunbergii	Fr	
vinifera	L, Fr	
Zygophyllaceae		
Tribulus		
terrestris	Se	

Appendix 2

Botanic Gardens and Arboreta Across China

Province	Title	Year of Foundation	Location
Auhui	Hefei Bot. Gard.	1987	Hefei
Beijing	Beijing Medicinal Plant Gard.	1984	Inst. of Med. Plants of Chinese Acad. Med. Sci.
	Beijing Bot. Gard. (North)	1955	Xiangshan, Beijing
	Beijing Bot. Gard. (South)	1955	Xiangshan, Beijing
Fujian	Fuzhou Arb.		Xindian Town, Fuzhou
	Xiamen Bot. Gard.	1960	Xiamen
Gansu	Minquin Psammophyte Gard.	1974	Minquin County
Guangdong	Shenzhen Xianhu Bot. Gard.	1982	Liantang, Shenzhen
	South China Bot. Gard.	1959	N. E. Guanzhou
Guangxi	Guangxi Medicinal Plant Gard.	1959	Maoqiao, Nanning
	Guilin Bot. Gard.	1958	Yanshan, Guilin
	Nanning Arb.	1979	Youi Road, Nanning
Guizhou	Guiyang Medicinal Plant Gard.	1984	Shachong Road, Guiyang
	Guizhou Bot. Gard.	1964	Luchongguan
Hainan	Arboretum, Hainan Inst. of Forest Sci.	1960	Fengmu Town, Tunchang
	Hainan Tropical Economic Plant Gard.	1958	Danxian County
Hebei	Wuhan Bot. Gard.	1956	Donghu, Wuchang
Heilongjiang	Heilongjiang Forest & Plant Gard.	1958	Harbin
Hong Kong AR	Hong Kong Bot. Gard.	1861	Albany Road, Hong Kong
Hunan	Hunan Forest Plant Gard.	1985	S. Changsha
	Nanye Arb.	1978	Hengyang
Inner Mongolia	Dengkou Psammophyte Gard.	1986	Dengkou County
	Hohhot Arb.	1956	Inner Mongolia Acad. Forest Sci.
Jiangxi	Gannan Arb.	1976	Doushi Town, Shangyou

	Lushan Bot. Gard.	1934	Lushan Mountain
	Nanjing Medicinal Plant Gard.	1958	Chinese Medicine Coll. China Univ. of Pharmaceutics, N. Nanjing
	Nanjing Zhongshan Bot. Gard.	1929	E. Nanjing
Jilin	Changchun Forest Plant Gard.	1982	Jingyuetan National Forest Park, Changchun
	Junjiang Arb.	1985	Hunjiang Inst. of Forest Sci.
	Hunjiang Arb.	1985	Hunjiang Inst. of Forest Sci.
Liaoning	Bot. Gard. Of Shenyang Inst. Applied Ecol.	1955	Inst. App. Ecol. Shenyang
	Shenyang Arb.		Pingniandajie Xiongyue Town, Shenyang
	Xiongyue Arb.	1915	Xiongyue Town, Shenyang
Ningxia	Yanchi Xeric Sand Plant Gard.	1984	Nanmemwai, Yinchuan
	Yinchuan Bot. Gard.	1986	S. Yinchuan
Qinghai	Xining Bot. Gard.	1981	Xinning Road, Xining
Shaanxi	Xi'an Bot. Gard.	1959	South Cuihuanan Road, Xi'an
	Yanan Arb.	1980	Yangjiawan, Yanan
Shandong	Arb. of Shandong Forestry Sch.	1956	Deiguan, Taian
	Jinan Bot. Gard.	1986	S. Jinan
	Qingdao Bot. Gard.	1984	Qingdao
Shanghai A. R.	Shanghai Bot. Gard.	1954	Longhua
Shanxi	Wutaishan Mtn. Arb.	1988	Wutaishan Mtn.
Sichuan	Chengdu Bot. Gard.	1983	N. Chengdu
	Chongqing Flower Gard.	1988	Longxi Town Jiangbei
	Sichuan Medicinal Plant Gard.	1947	Sanquan, Nanchuan
Yunnan	Kunming Bot. Gard.	1938	Heilongtan, Kunming
	Kunming Bot. Gard. Of Garden Plants	1978	E. Kinming
	Xishuangbanna Tropical Bot. Gard.	1959	Menglun, Mengla
Xinjiang	Tulufan Desert Plant Gard.	1976	Tulufan City
	Urumqi Bot. Gard.	1984	Urumqi
Zhejiang	Bot. Gard. of Zhejang Agric. Univ.	1927	Huajiachi, Hangzhou
	Hangzhou Bot. Gard.	1984	Yuguan, Hangzhou
	Wenzhou Bot. Gard.	1966	Shejiang Inst. of Tropical Plants
	Zhejiang Bamboo Gard.	1982	W. Hangzhou

The above table contains more than 50botanic gardens and arboreta but others could be added. The reader will recall the Kadoorie Centre in Hong Kong and the Huaxi sub-alpine garden in Sichuan raising the number further. Additionally, there is of course a botanic garden in Taipei.

Once in China and prepared to enquire, eventually it is possible to find little publicised specialist collections in various institutions and the list given here is not exhaustive. It does, however, offer a way into Chinese plant life on a province by province basis. Although gardens and arboreta concentrate a great deal in a relatively small space it is, in the interests of a wider understanding, important, too, to appreciate the complementary significance of the increasing number of reserves becoming available in China and to some of which this book has made reference.

Data modified from *Guide Book for Visitors to Botanical Gardens of China* 1991 and augmented.

Appendix 3

Chinese Dynasties

Neolithic ca. 5000 – ca. 1700 B.C.

Shang ca. 1700 – ca. 1050 B.C.

Zhou, ca. 1050 – 221 B.C.

Qin 221 – 206 B.C.

Han 206 B.C. – A.D. 220

Six Dynasties 220 – 589

Sui 581 – 618

Tang 618 – 907

Five Dynasties 907 – 960

Song 960 – 1127

Yuan Dynasty 1270 – 1368

Ming Dynasty 1368 – 1644

Qing Dynasty 1644 – 1911

Republic of China 1911 – 1949

People's Republic of China 1949 –

References

Allchin, F R (1969) Early cultivated plants in India and Pakistan. In Ucko P J and Dimbleby G W (Eds.) *The Domestication and Exploitation of Plants and Animals*

Duckworth, London

Andrus J R and Mohamed A F (1958) *The Economy of Pakistan*

Oxford University Press

Anon (1996) 'Neither food nor medicine

'Which' magazine September pp. 24–24

Anon (1998) 'When is a food not a food'

Chemist and Druggist 25th July p 5

Anon (2000) *Flora of Hong Kong.* Pre-publication data

c/o Corlett, R (University of Hong Kong)

Arber, A (1986) *Herbals. Their Origin and Evolution. A Chapter in the History of Botany 1470–1670*

(Re-issue of 2nd Edn. 1938) Cambridge University Press

Argent G, Mendum M and Smith P (1999) The smallest Rhododendron in the world, *R. caespitosum*

New Plantsman 6 152–157

Attele A S, Wu J A and Yuan C S (1999) Ginseng pharmacology: multiple constituents and multiple actions

Biochem. Pharmacol. 58 1685–93

Austin D (1988) *The Heritage of the Rose*

Antique Collector's Club, Suffolk

Axelrod D I, Al-Shehbaz I and Raven P H (1998 History of the modern flora of China. In Zhang A L and Wu S G (Eds.) *Floristic Characteristics and Diversity of East Asian Plants*

China Higher Education Press. Springer-Verlag, Hong Kong

Babock E B (1943) Systematics, cytogenetics and evolution in *Crepis*

Bot. Rev. 8 139–180

Balz J P, Courtois D, Drieu J, Drieu K, Reynoird J P, Sohier C, Teng B P, Touche A and Petiard V (1999) Production of ginkgolides and bilobides by *gingko biloba* plants and tissue cultures

Planta. Med. 65 620–626

Bartholomew B, Boufford D E and Spongberg S A (1983) *Metasequoia glyptostroboides* – its present status in central China

J Arnold Arbor. 64 105–128

Bayard D T (1970) Excavation of Non Nok Tha, northeastern Thailand, 1968
Asian Perspective **13** 109–143

Beakman A C, Wierenga P K, Woerdenbag H J, van-Uden W, Pras N, Konings A W, El-Feraly F S, Galal A M and Wilkstrom H V (1998) Artemisin – derived sesquiterpene lactones as potential anti-tumour compounds: cytotoxic action against bone marrow and tumour cells.
Planta. Med. **64** 615–619

Beal J L and Reinhard E (1980) *Natural Products as Medicinal Agents*
Pub. Hippokrates

Beales P (1997) *Classic Roses: An illustrated encyclopaedia and growers manual of old roses, shrubs and climbers*
Harvill Press, London

Beckett K and Grey-Wilson C (1993, 1994) *Encyclopaedia of Alpines* (volumes 1 and 2 in successive years)
Alpine Garden Society Publications, Pershore, UK

Bennet D, Maglio C M and Martin S (1995) Americas first Camellias
Amer. Camellia Year Book **48**–50

Bentham G (1861) *Flora Hongkongensis*
Lovell Reeve, London

Bessey C E (1915) The phylogenetic taxonomy of flowering plants
Ann. Missouri Bot. Gard. **2** 109–164

Blasco G and Condell G A (1988) Recent developments in the chemistry of plant derived anti-cancer agents
Econ. And Medicine Plant Res. **3** 119–191

Bray F (1984) Agriculture 1–724. In Needham J (Ed) *Science and Civilisation in China*
Vol. 6, Part 2. Cambridge University Press

Bretschneider E (1898) History of European Botanical Discoveries in China
Sampson, Low, Marston and Co. London

Chaney R W (1948) The bearing of the living *Metasequoia* on problems of Tertiary paleobotany
Proc. Natl. Acad. USA **34** 503–515

Chang D H S (1981) The vegetation zonation of the Tibetan plateau
Mountain Res. and Devel. **1** 29–48

Chang D H S (1983) The Tibetan plateau in relation to the vegetation of China
Ann. Missouri Bot. Gard. **70** 564–570

Chang H T (1981) A taxonomy of the genus *Camellia*
Acta Seientarium Naturalium Universitas Sun Yat Sen 1–180

Chang T T (1976) The origin, evolution, cultivation, dissemination and diversification of Asian and African rices
Euphytica **25** 425–441

Chapman G P (1997) *The Bamboos* Linnean Society Symposium Series 19
Acad. Press. London

Chen F P (1986) A study of the limestone flora of Longgang (south western Guangxi).
Masters Thesis, Zhongshan University, Guangzhou

Chen Z K, Zhang J H and Zhou F (1995) The ovule structure and development of the female gametophyte in *Cathaya* (Pinaceae)

Cathaya **7** 165–176

Chun W-y and Kuang K-z (1962) De genera *Cathaya* Chun et Kuang

Acta Bot. Sin. **10** 245– 246

Clayton W D and Renvoize S A (1986) *Genera Graminum: Grasses of the World* Kew Bull. Add. Ser. 13. Royal Botanic Gardens, Kew, London

Cox P and Cox K (1997) Encyclopaedia of Rhododendron species

Glendoick Publishers, Glencarse

Coates A M (1969) *The Quest for Plants. A History of the Horticultural Explorers* Pub. Studio Vista London

Cui J, Garle M, Eneroth P and Bjorkhem J (1994) What do commercial ginseng preparations contain?

Lancet **344** 134

Cullen J, Alexander J C M, Brickell C D, Edmondson J R, Green P S, Heywood V H, Jørgensen P-M, Jury S L, Knees S G, Matthews V A, Maxwell H S, Miller D M, Nelson E C, Robson N K B, Walters S M and Yeo P F. *The European Garden Flora Vol. V Dicotyledons Part III Limnanthaceae to Oleaceae*

Cambridge University Press

Curtis-Prior P, Vere D, Robbins t and Fray P (1998) Therapeutic Value of *Ginkgo* in reducing symptoms of decline in mental function

Pharmacy and Pharmacology **50** (suppl.) 19

Dahlgren R M T, Clifford H T and Yeo P F G (1985) *The Families of Monocotyledons. Structure, Evolution and Taxonomy*

Springer-Verlag Berlin

Daniels C (1996) Agro-industries: sugar cane technology pp. 1–539 in Needhams J *Science and Civilisation in China* Vol. 6 Part III

Cambridge University Press

Darlington C D and Wylie A P (1955) *Chromosome Atlas of Cultivated Plants*

George, Allen and Unwin, London

Darwin C (1868) *Animals and Plants under Domestication* 1st Edn. 2 vols.

John Murray, London

Debener T, Bartels C and Mattiesch L (1996) RAPD analysis of genetic variation between a group of rose cultivars and selected wild rose species

Molec. Breeding **2** 321–327

Debener T and Mattiesch L (1999) Construction of a genetic linkage map for roses using RAPD and AFLP markers

Theor. Appl. Genet. **99** 891–899

Dore S, Bastianetto S, Kar S and Quirion R (1999) Protective and rescuing abilities of IGF – I and some putative free radical scavengers against beta-amyloid inducing toxicity in neurons

Ann. New York Acad. Sci. **891** 356–364

Eliot G (1871) *Middlemarch*

Reprinted 1994, Penguin Books, London

Feng K (1999) *Rare and Precious Wild Flowers of China*

China Forestry Publishing House, Beijing

Ferguson D K (1967) On the phytogeography of Coniferales in the European Ce-
nozoic
Palaeogeogr. Palaeoclimatol and Palaeoecol. **3** 73–110

Filgar R B (1993) Stone Magnolias
Mag. of the Arnold Arbor. **53** 3–9

Florin R (1963) The distribution of conifer and taxad genera in time and space
Acta. Hort. Berg. **20** 121–312

Foster A S (1962) Ontogenetic studies of dichotomous venation in *Kingdonia uniflora* (Abst.)
Amer. J. Bot **49** 659–321

Foster A S (1963) The morphology and relationships of *Circeaster*
J Arnold Arbor **44** 299–321 + 5 plates

Foster A S (1966) Morphology of anastomoses in the dichotomous venation of
Circeaster
Amer. J Bot. **53** 588–599

Foster A S and Arnott H J (1960) Morphology and dichotomous vasculature of the
leaf of *Kingdonia uniflora*
Amer. J. Bot. **47** 684–698

Fretwell B (1989) *A Comprehensive Guide to Clematis*
Harper Collins

Fu L-K and Jin J-M, (1992) (Eds.) *China Plant Red Data Book Vol. 1*
Science Press, Beijing

Gajewski W (1946) Cytogenetic investigations on *Anemone* L. 1. *Anemone
jancjewskii*, a new amphidiploid species of hybrid origin
Acta. Soc. Bot. Polon. **17** 129–194 (quoted by Stebbins G L 1950)

Gandolin M H, Reynders-Aloisi S, Mando B, Debener T, Drewes-Alvarez R,
Spellerberg B, Cubero J and Roberts A (2000) European network for characeri-
sation and evaluation of the genus *Rosa* germ plasm.
Acta Horticulturae No. 508 341–344

Ge S, Hong D Y, Wang H Q, Liu Z Y and Zhang C H (1998) Population genetic
structures and conservation of an endangered conifer *Cathaya argyrohylla*
(Pinaceae)
Int. J. Plant Sci **159** 351–357

Ge S, Wang H Q, Zhang C M and Hong d Y (1997) Genetic diversity and population differentiation of *Cathaya argyrophylla* in Bamian Mountain
Acta. Bot. Sin **39** 266–271

Good R (1925) The past and present distribution of the Magnolieae
Ann. Bot. **39** 409–430

Good R (1964) *The Geography of The Flowering Plants*
Longman

Gou C H (1989) *History of Ancient Biology in China*
Science Press, Beijing

Gregus P (1972) *Xylotomy of the living Conifers*
Academiai Kiado, Budapest

Grey-Wilson C (2000) *Clematis, the Genus* (A Comprehensive Guide for Gardeners, Horticulturalists and Botanists)
Batsford

Grist D H (1951) *Rice*
Longman

Gu Z, Lifang X, Lishan X and Kondo X (1989) Report of the chromosome numbers of some species of *Camellia* in China
Amer. Camellia Year Book 19–22

Halda J J (1996) *The Genus Gentiana*
Dobre. Czech Republic

Hansen C (1990) New species and combinations in *Allomorphia, Phyllagathis* and *Sporexia* (Melestomotaceae) in Indo-China
Bull. Mus. Natl. Hist. Nat. B. Adamsonia 12 37–41

Hantula J, Notila P, Saura A and Lokki J (1989) Chloroplast DNA variation in *Anemone s. lato* (Ranunculaceae)
Pl. Syst. Evol. 163 81–85

Hardt R W and Capule C (1983) Adoption, spread and production impact of modern rice varieties in Asia
IRRI. Philippines

Haudricourrt A G and Metailie G (1994) De L'illustration botanique en Chine
Etudes Chinoises 13 381–416

Haw S G (1996) Notes on some Chinese and Himalayan Rose species of Section Pimpinellifoliae
The New Plantsman 3 143–146

Ho T-N (1985) A study on the genus *Gentiana* of China
Acta. Phytotaxonomica Sinica 23 43–52

Ho T-N (1988) Gentianaceae
Flora Reipublicae Popularis Sinica Vol. 62. Science Press, Beijing

Ho T-N and Liu S W (1990) The infrageneric classification of *Gentiana*, (Gentianaceae)
Bull. Brit. Mus. (Nat, Hist.) Bot. 20 169–192

Hoot S B (1995) Phytogenetic relationships in *Anemone* based on DNA restriction site variation and morphology pp. 295–300 In Jensen U and Kadereit J W (Eds.) *Systematics and Evolution of the Ranunculaceae*
Springer, New York

Hoot S B and Palmer J D (1994) Structural rearrangements including parallel inversions within the chloroplast genome of *Anemone* and related genera
J Mol. Evol. 38 274–281

Hoot S B, Raznicets A A and Palmer J D (1994) Phylogenetic relationships in *Anemone* (Ranunculaceae) based on morphology and chloroplast DNA
Syst. Bot. 19 169–200

Hu H-H (1946) Notes on a Palaeogene species of *Metasequoia* in China
Bull. Geol. Soc. China 26 105–107

Hu Y X and Wang F X (1984) Anatomical studies of *Cathaya* (Pinaceae)
Amer. J. Bot. 71 727–735

Huang Y and Chen J (1988) Sea level changes along the coast of the south China sea since the late Pleistocene pp. 289–318 in white P (Ed.) *The Palaeoenvironment of East Asia from the Mid Tertiary*
Pub. Centre of Asian Studies. Univ. of Hong Kong

Hurst C C (1941) Notes on the origin and evolution of our garden roses
Jour. Roy. Hort. Soc. **66** 73–82, 242–250, 282–289

Israelson L D (1997) Harmonising North American herbal regulation: a US perspective
Herbal Gram. **32** 20–21

Junell S (1931) Die entwicklungsgeschichte von *Circeaster agrestis*
Svensk Botanisk Tidskrift **25** 238–270

Keng H, Hong D-Y and Chen C-J (1993) *Orders and Families of Seed Plants of China*
Pub. World Scientific Publishing Co. Singapore

Kenyon K H (1978) *Archaeology in the Holy Land*
4th Edn. Ernest Benn Ltd., London

Klayman D L (1993) *Artemisia annua* From seed to respectable antimalarial plant pp. 242–285 in Kinghorn A D and Balandrin M F (Eds.) *Human Medicinal Agents from Plants*
Amer. Chem. Soc. Symp. Ser. 534

Kleijuen J and Knipschild P (1992) *Ginkgo biloba* for cerebral insufficiency
Br. J. Clin. Pharmacol. **34** 352–358

Kong H-Z and Yang Q (1997) Karyomorphology and relationships of the genus *Circeaster* Maxim
Acta Phytotax. Sin. **35** 494–499

Kong S C, Lui C J and Zhang J Z (1996) Discovery of rice remains in the Jia lake site, Wuyang county, Henan province and their significance in the archaeology of the human environment.
Archaeology **12** 78–85

Küng H (1987) *Christianity and the World Religions*
Collins Fount Paperback, London

Kurashige Y, Mine M, Eto J, Kobayashi N, Handa T, Takjayanagi K and Yukawa T (2000) Sectional relationships in the genus *Rhododendron* (Ericaceae) based on *matK* sequences p. 349 in: Andrews S, Leslie A and Alexander C (Eds.) *Taxonomy of Cultivated Plants*
3rd Int. Symp. Royal Botanic Gardens, Kew, London

Lancaster C R (1989) *Travels in China, A Plantsman's Paradise*
Antique Collector's Club, Woodbridge, Suffolk, UK

Lang K, Feng Z and Li B (1997) *Alpine Flowering Plants in China*
Yunnan Acad. For Sci. Esperanto Press, Beijing

Li H and Zheng Y (1995) Palaeogene floras pp. 455–505 in Li X (Ed.) *Fossil Floras of China Through the Geological Ages*
Guangdong Science and Technology Press, Guangzhou, China

Li X W and Li J (1993) A preliminary floristic study on the seed plants from the region of Hengduan Mountains
Acta. Bot. Yunnan **15** 217–231

Li Y (1996) *List of Plants in Xishuangbanna*
C A S Xishuangbanna Trop. Bot. Gard. and Dept. of Ethnobot. Kunming Inst. Bot.
Liang C F, Liang J Y, Lau L F and Mo X L (1985) A report on the exploration of
the flora of Longgang (southwestern Guangxi)
Guihaia **5** 191–209
Liang C F, Su Z Z M and Zhou K J (1981) A karst forest reservation in north
tropical Guangxi
Guihaia **1** 1–6
Lin L K, Lin Y R and Zhang Y T (1981a) A list of vascular plants from Wuyis-
han, north Fujian
Wuyi. Science J. Suppl. **1** 17–69
Lin Y R, Wang X W and Zhang G C (1981b) A preliminary study of the flora of
the spermatophyta in the Wuyishan nature reserve WYNR in Fujian
Wuyi Science Journal Suppl. **1** 57–82
Lord T (1997-8) (Ed.) *The RHS Plant Finder 1997–98*
Dorling Kindersley. London, New York, Sydney, Moscow
Lu G-D and Hueng H-T (1986) Botany. In Needham J (Ed) *Science and Civilisa-
tion in China*
Vol 6 Part 1 Cambridge University Press
Lu T and Huang L (1995) Yellow camellias in Guangxi
Int. Camell. J. **27** 82–83
Lyte C (1989) *Frank Kingdon-Ward, The Last of the Great Plant Hunters*
John Murray, London
Maheshwari P (1950) an *Introduction to the Embryology of Angiosperms*
McGraw Hill, New York
Martin M, Piola F, Chessel D, Jay M and Heizman P (2001) The domestication
process of the modern rose: genetic structure and allelic composition of the rose
complex.
Theor. Appl. Genet. **102** 398–404
Matsumoto S, Kouchi M, Yabuki J, Kusunoki M, Ueda Y and Fukui H (1998)
Phytogenetic analyses of the genus *Rosa* using the *mat*K sequence: molecular evi-
dence for the narrow genetic background of modern roses
Scientia Hortic. **77** 73–82
Menzies N K (1996) Forestry in Needham J (Ed.) *Science and Civilisation in
China*
Vol. 6 Part III 541–740
Metailié G (1995) Le travail de la citation en Chine et au Japon
Extreme Orient: Extreme Occident **17** 131–139, Univ. de Vincennes
Miki S (1941) On the change of flora in eastern Asia since the Tertiary Period. 1.
The clay or lignite beds flora in Japan with special reference to the *Pinus* trifo-
lia beds in central Hondo
Japan. J. Bot. **11** 237–303
Mills S Y (1996) Chinese herbs in the West pp. 505–510 In: Trease G E and Ev-
ans W C (Eds.) *Pharmacognosy* 14ᵗʰ Edn.
W B Saunders, London

Mitchell A F (1970) Recent measurements of *Metasequoia* in Britain
J. Roy. Hort. Soc. **45** 452

Mizota C, Endo H, Um K T, Kusakabe M, Noto M and Matsuhisa Y (1991) The colian origin of silty mantle in sedimentary soils from Korea and Japan
Geoderma **49** 153–164

Moser E (1991) Rhododendren
Neuman – Verlag

Muller J (1970) Palynological evidence on early differentiation of angiosperms
Biol. Revs. **345** 417–450

Needham J, Lu G-D and Huang H-T (1986) Botany pp. 1–78 in Needham J (Ed.). *Science and Civilisation in China* Vol. 6 Part I
Cambridge University Press

Ogisu M (1996) Some thoughts on the history of China roses
The New Plantsman **3** 152–157

Oka H I (1988) *Origin of Cultivated Rice*
Elsevier

Olmstead R G and Palmer J D (1994) Chloroplast DNA and systematics–a review of methods and data analysis
Amer. J Bot. **81** 1205–1224

Oxelman B and Lidén M (1995) The position of *Circeaster*–evidence from nuclear ribosomal DNA pp. 189–194 In (Eds.) Jensen U and Kadereit J W *Plant Syst. Evol. Suppl.* **9** Ranunculiflorae

Page M (1997) *The Gardeners Guide to Growing Peonies*
David and Charles, Newton Abbot Timber Press, Portland, Oregon

Palmer J D, Jansen R K, Michaels H J, Chase M W and Manhart H R (1988) Chloroplast DNA variation and plant phylogeny
Ann. Missouri Bot. Gard. **75** 1180–1206

Parks C R (1992) Classification of *Camellia* species: new approaches
Amer. Camellia Year Bk. 13–21

Parks C R (1995) Inspiration from China for the *Camellia* breeder
Amer. Camellia Year Bk. 80–83

Parks C R, Yoshikawa N, Prince L and Thakor B (1995) the application of *isozymic* and molecular evidence to taxonomic and breeding problems in the genus *Camellia*
Int. Camellia J **27** 103–111

Petit R J, Pineau E, Demesure B, Bacilieri R, Ducousso A and Kremer A (1997) Chloroplast DNA footprints of post glacial recolonisation by oaks
Proc. Natl. Acad. Sci. USA **94** 9996–10,001

Phillips R and Rix M (1993) *The Quest for the Rose*
BBC Books, London

Phillipson J D (1994) Traditional medicine treatment for eczema: experience as a basis for scientific acceptance
Eur. Phytotel No. 6 August 33–40

Phillipson J D (1995) A matter of some sensitivity
Phytochem **38** 1319–1343

Philo 'Works' Translated by Yonge C D (1993)

Hendrickson USA

Postan C (1996) (Ed.) *the Rhododendron Story: 2,000 years of Plant Hunting and Garden Collection*

Roy. Hort. Soc., London

Prince L M and Parks C R (1997) Evolutionary relationships in the tea subfamily Theoideae based on DNA sequence data

Int. Camellia J. **29** 135–144

Read B E (1946) Famine foods listed in the *Chui Huang Pen T'shao*

Lester Institute, Shanghai

Read D P (1993) *Chinese Herbal Medicine*

Shambala

Reid E M and Chandler M E J (1933) *The London Clay Flora*

Br. Mus. London

Ren Y and Hu Z (1996) Morphological studies on anastomoses and blind veins in dichotomous venation of the leaf in *Kingdonia uniflora*

Acta Phytotax. Sin. 34

Ren Y, Hu Z and Li Z-j (1997) the morphology of the dichotomous leaf venation of *Circeaster* and its systematic implication

Acta Phytotax. Sin. **34** 569–576

Ren Y and Hu Z (1998a) Anatomical studies on root, node and leaf of *Kingdonia uniflora*

Acta Bot. Boreal-Occident Sin **18** 72–72

Ren Y and Hu Z (1998b) The discovery of the bilacunar nodes and bilobed leaves in *Circeaster* and their significance

Acta Bot. Boreal-Occident Sin. **18** 566–569

Ren Y, Wang M-L and Hu Z-H (1998) *Kingdonia*, embryology and its systematic significance

Acta Phytotaxonomica sin **36** 423–427

Ren Y, Xiao Y-P and Hu Z-H (1998) The morphological nature of the open dichotomous leaf venation of *Kingdonia* and *Circeaster* and its systematic implication

J. Plant Res. **III** 225–230

Rigney U, Kimber s and Hindmarch I (1999) The effects of acute doses of standardised *Ginkgo biloba* extract on memory and psychomotor performance of volunteers

Phytotherapy Res. **13** 408–415

Roberts A V (1977) Relationship between species in the genus *Rosa*, section Pimpinellifoliae

Bot. J. Linn. Soc. **74** 309 – 328

Roberts A V, Blake P S, Lewis R, Taylor J M and Dunstan D J (1999) The Effects of Gibberellins on flowering in Roses

J. Plant Growth Regul. **18** 113–119

Salley H E and Greer H E (1986) *Rhododendron Hybrids: a Guide to their Origins*

Timber Press, Portland, Oregon

Sang T, Crawford D J and Stuessy T F (1995) Documentation of reticulate evolution in peonies (*Paeonia*) using sequences of internal transcribed spacer of nuclear ribosomal DNA: implications for biogeography and concerted evolution
Proc. Natl. Acad. Sci. USA **92** 6813–6817

Sang T, Crawford D J and Stuessy T F (1997) Chloroplast DNA phylogeny, reticulate evolution and biogeography of *Paeonia* (Paeoniaceae)
Amer. J. Bot. **84** 1120–1136

Sang T and Zhang D (1999) Reconstructing hybrid speciation using sequences of low copy nuclear genes: hybrid origins of five *Paeonia* species based on *Adh* gene phylogenies
Syst. Bot. **24** 148–163

Savidge T (1994) The first yellow *Camellias*
Int. Camellia J. **26** 78–80

Sealy J R (1958) *A Revision of the Genus Camellia* VIII
Roy. Hort. Soc. London

Seward A C (1938) The story of the Maidenhair tree
Science Progress **32** 420–440

Shoyama Y, Matsumoto M and Nishioka I (1987) Phenolic glycosides from diseased roots of *Rehmannia glutinosa* var. *purpurea*
Phytochem. **26** 983–986

Siu Lai-Ping G (2000) Orchidaceae pp. 132–141, *Flora of Hong Kong*

Smith B S (1994) *The Emergence of Agriculture*
Sci. Amer.

Spencer J E (1963) the migration of rice from mainland south east Asia into Indonesia pp. 83–89 In: Barrau J (Ed.) *Plants and the Migration of Pacific Peoples*
Bishop Museum Press, Honolulu

Stearn W T (1990) Review of Lyte C Kingdon Ward
Bot. J. Linn. Soc. **104** 397–399

Stearn W T (2000) Early introduction of plants from Japan into European gardens pp. 337–340 In:; Andrews S, Leslie A and Alexander C (Eds.) *Taxonomy of Cultivated Plants*
3rd Int. Symp. Royal Botanic Gardens, Kew, London

Stebbins G L (1950) *Variation and Evolution in Plants*
Columbia University Press, New York

Takeuchi S, Bomura K, Uchiyama H and Youeda K (2000) Phylogenetic relationships of the genus *Rosa* based on the restriction enzyme analysis of the chloraplast DNA
J. Japan. Soc. Hort. Sci. **69** 598–64

Takhtajan A (1969) *Flowering Plants: Origin and Dispersal* Translated by C. Jeffrey
Oliver and Boyd, Edinburgh

Tang W and Eisenbrand G (1992) *Chinese Drugs of Plant Origin*
Springer-Verlag

Thakor B H (1997) A re-examination of the phylogenatic relationships within the genus *Camellia*
Int. Camellia J. **29** 130–134

Theophrastus *Enquiry into Plants*
trans. Sir Arthur Hort. 1916
Van Gelderen D M and van Hoey Smith J R P (1992) *Rhododendrons* Translated from the Dutch by Handgraaf N and T
Batsford London. Royal Boskoop Hort. Soc.
Vavilov N I (1920) The Origin, variation, immunity and breeding of cultivated plants. Translated by K Starr Chester
Chron. Bot. **13** 1–364
Vishnu-Mittre (1974) Paleobotanical evidence in Indian rice. In Hutchinson J B (Ed.) *Evolutionary Studies on World Crops*
Cambridge University Press
Wall M E and Wani M C (1993) Camptothecin and analogues: synthesis, biological in vitro and in vivo activities pp. 149–169, In: Kinghorn A D and Balandrin M F (Eds.) *Human Medicinal Agents from Plants*
Amer. Chem. Soc. Symp. Ser. **534** Washington
Wang D-y and Liu H-z (1982) A new species and a new variety of *Cunninghamia* from Sichuan Province
Acta. Phytotaxonomica Sin. **20** 230–232
Wang F X (1990) *Biology of the Cathaya* [in Chinese]
Science Press, Beijing
Wang F X and Chen Z K (1974) The embryogeny of *Cathaya* (Pinaceae)
Acta Bot. Sin. **16** 64–69
Wang K (1990) Ethnobotanical studies of bamboo resources in Mensong, Xishuangbanna, Yunnan, China pp. 1–7 In: 2nd Int. Cong. Ethnobotany, Kunming, Yunnan, China
Wang K, Ding M, Yin Z and Liu H (1988) the Loess in the middle reaches of the Yellow River in China and its palaeographic environment pp. 433–444. In: White P (Ed.) *the Palaeoenvironment of East Asia from the Mid Tertiary*
Centre for Asian Studies, University of Hong Kong
Wang K, Xue J C S and Pei S A K (1993) Ethnobotanical studies of bamboo in Xishuangbanna, Yunnan, China
Papers of Tropical Botany Resources, Yunnan University **8** 47–65
Wang L (undated) A glance over the flora of Xinglong Mountain in Gansu, China
Occasional paper, Dept. Biology, Northwest Normal University, Lanzhou
Wang W T (1980) Ranunculaceae
Flora Reipublicae Popularis. Sin. Vol. 28, Science Press, Beijing
Wang W T (1998) Notulae de Ranunculaceis Sinensibus (xxii)
Acta Phytotaxonomica Sin. **36** 156–172
Wang W Ts and Wu S G (1993) *Vascular Plants of the Hengduan Mountains Vol. 1*
Science Press, Beijing
Wang W Ts and Wu S G (1994) *Vascular Plants of the Hengduan Mountains Vol. 2*
Science Press, Beijing
Wang X Q, David C T and Sang T (2000) Phylogeny and divergence times in the Pinaceae: evidence from three genomes

Mol. Biol and *Evol* **17** 773–781
Wang X Q, Han Y, Deng Z R and Hong D Y (1997) Phylogeny of the Pinaceae evidenced by molecular biology

Acta Phytotax. Sin **35** (2) 1–10
Wang X Q, Han Y and Hong D Y (1998) A molecular systematic study of *Cathaya* a relict genus of the Pinaceae in China

Pl. Syst. Evol. **213** 165–172
Wang X Q, Zhou Y P, Zhaang d M and Hong D Y (1997) Genetic diversity analysis by RAPD in *Cathaya argyrophylla* Chun. Et Kuang

Science in China (Series C) **40** (2) 91–97
Wang Z (1992) The little ice age of the north west region of China

China Geog. Sci. **2** 215–225
Wang Z C (1991) *Ancient Biology of China*

Educational Press of Shandong
Wang Z R (1994) *Botanical History of China*

Science Press, Beijing
Wendel J F and Parks C R (1985) Genetic diversity and population structure in *Camellia japonica* . (Theaceae)

Amer. J. Bot. **72** 52–65
Wilkes G (1990) Quoted in Fowler C and Mooney P *The Threatened Gene: Food, Politics and the Loss of Genetic Diversity*

Lutterworth Press. Cambridge
Williamson E M, Okpako D T and Evans F J (1996) *Selection, Preparation and Pharmacological Evaluation of Plant Material, vol. 1*

Wiley
Wilson E H W (1929) *China, Mother of Gardens*

Boston 1929, reprinted 1971 by Benjamin Blom. New York
Wu S, Ueda Y, Nishihara S and Matsumoto S (2001) Phylogenetic analysis of Japanese *Rosa* species using DNA sequences of nuclear ribosomal internal transcribed spacers (ITs)

J Hort. Sci. and Biotech. **76** 127–32
Wu Z and Raven P H (1994) Flora of China Vol. 17
Wu Z and Raven P H (1996) Flora of China Vol. 15
Wu Z and Raven P H (1998) Flora of China Vol. 18
Wu Z and Raven P H (1999) Flora of China Vol. 4
Wu Z and Raven P H (1999) Flora of China Vol. 16

All volumes: Missouri Botanic Garden Press, St Louis
Science Press, Beijing
Wu Z Y (1988) Hengduan mountain flora and its significance

Japan. J. Bot **63** 297–311
Wu Z Y and Wu S G (1998) A proposal for a new floristic kingdom (realm) – the E. Asian kingdom, its delineation and characteristics pp.   In: Zhang A L and Wu S G (Eds.) *Floristic Characteristics and Diversity of East Asian Plants*

China Higher Education Press, Beijing
Springer-Verlag, Hong Kong

Wylie A P (1954, 1955) The history of garden roses. (Master's Memorial Lecture) and published as

Part I (1954) *J. Roy. Hort. Soc.* **79** 555–571
Part II (1955) *J. Roy. Hort. Soc.* **80** 8–24
Part III (1955) *J. Roy. Hort. Soc.* **80** 77–87.

Wyman A P (1968) *Metasequoia* after 20 years in cultivation

Arnoldia **28** 113–123

Xia L (1996) An outline of the studies on *Camellia* in Kunming Institute of Botany

Int. Camellia J. **28** 58–61

Xia W Y (1981) *Explanation of the Agricultural Affairs in 'The Book of Songs'*

Agricultural Press, Beijing

Xiao P-G (1994) Ethnopharmacological investigation of Chinese medicinal plants pp. 167–177. In: Chadisich D J and Marsh J *Ethnobotany and the Search for New Drugs*

CIBA Foundation

Xie Z Q (1996) Studies on the population ecology of *Cathaya argyrophylla*

PhD Diss. CAS. Institute of Botany, Beijing

Xie Z Q and Chen W L (1994) The present status and the future of *Cathaya argyrophylla*

Forest. China. Biodiv. **2** 11–15

Xu Q (1988) Southward migration events of mammals in east Asia during the Pleistocene and their relations with climatic changes pp. 873–882. In:; White P (Ed.) *The Palaeoenvironment of East Asia from the Mid Tertiary Vol. 2*

Centre for Asian Studies, University of Hong Kong

Xu Z R (1986) An overview of the studies of Chinese limestone forests: problems and strategies

Ecologic. Science **2** 98–102

Xu Z R (1993a) A study of the limestone forest flora of southern and south western China

Guihaia (Suppl.) **4** 5–54

Xu Z R (1993b) A species list of limestone plants in China

Guihaia (Suppl.) **4** 155–258

Xu Z R (1995) A study of the vegetation and floristic affinity of the limestone forest in southern and south western China

Ann. Missouri Bot. Gard. **82** 570–580

Xu Z R and Sun L (1984) A preliminary study of the limestone hill forest vegetation and flora of Dongyang Mountain area (Southern Guizhou)

Ecologic. Science **2** 1–6

Xue D (1991) *Guidebook for Visitors to Botanical Gardens of China*

Jin Deng Publishing House, Beijing

Yang Z, Shang Q, Feng Z, Lang K and Li H (1993) *Orchids* Translated by Xiong Z R

Chinese Esperanto Press, Beijing

Ying T S, Zhang Y-L and Boufford D E (1993) *The Endemic Genera of Seed Plants of China*

Science Press, Beijing

Yoshikawa K and Yoshikawa N (1996) *Camellia protojaponica* Huzioka: fossilised *Camellia* species from the middle Miocene

The Camellia J. **51** 24–27

Yoshikawa K and Yoshikawa N (1997) The native Camellia japonica and C. rusticana in Japan

Amer. Camellia Year Book 11–22

Yu T-t (1985) Rosaceae.

Flora Reipublicae Popularis Sin. Science Press, Beijing

Yuan Y, Kupfer P and Doyle J J (1996) Infra generic phylogeny of the genus *Gentiana* (Gentianaceae) inferred from nucleotide sequences of the internal transcribed spacers (ITS) of nuclear ribosomal DNA

Amer. J. Bot. **83** 641–652

Zhang D and Sang T (1998) Physical mapping of ribosomal RNA genes in peonies (*Paeonia,* Paeoniaceae) by fluorescent in situ hybridisation: implications for phylogeny and concerted evolution

Amer. J. Bot. **86** 735 – 740

Zieslin N and Halevy A H (1976) Flower bud atrophy in Baccara roses (V) The effects of different growth substances on flowering.

Physiol Plant **37** 326–330

Zhang L and Dai X (1989) The Loess Plateau–its formation and evolution. pp. 2-17 (Eds.) Zhang L and Zeng S Internat. Field Workshop on Loess Geomorphology–Processes and Hazards

J. Langzhou Univ. (Nat. Sci.) Suppl.

Zhang Q, Feng Z and Yang Z (1992) *Rare Flowers and Unusual Trees – a Collection of Yunnan's Most Treasured Plants*

CAS Institute of Botany, Kunming
China Esperanto Press, Beijing

Zhang W (1998) (Ed.) *China's Biodiversity: a Country Study*

China Environmental Science Press, Beijing

Zhang Y (1988) Preliminary analysis of the quaternary zoogeography of China based on distributional phenomena among land vertebrates pp. 883–896. In: White P (Ed.) *The Palaeoenvironment of East Asia from the Mid Tertiary, Vol. 2*

Centre for Asian Studies, University of Hong Kong

Zhao X (1992) (Ed.) *The Palaeoclimate of China*

Geological Publishing House, Beijing

Zhou B N (1995) The chemistry and bioactivities of some natural products from Chinese herbs pp. 313–334. In: Hostettman K, Marston A, Maillard M and Hamburger M (Eds.) Phytochemistry of Plants used in Traditional Medicine

Oxford Science Publications

Zhou S L, Hong D-Y and Pan K-Y (1999) Pollination biology of *Paeonia jishanensis* T Hong and W Zhao (*Paeoniaceae*), with special emphasis on pollen and stigma biology

Bot. J. Linn. Soc. **130** 43–52

Zhu T (1999) *Alpine Plants on the Changbashan Massif of China*
Science Press, Beijing and New York
Zieslin N and Halevy A H (1976) Flower bud atrophy in Baccara roses (v) the effects of different growth substances on flowering
Physiol. Plant **37** 326–330

Index of Chinese Classical Texts

For further details than appear in the present book the reader is referred to Needham J (Ed) *Science and Civilisation in China Volume 6.*

Taxonomic Index

Author Index

Subject Index

Note: for convenience all geographical areas are entered under 'regions'.

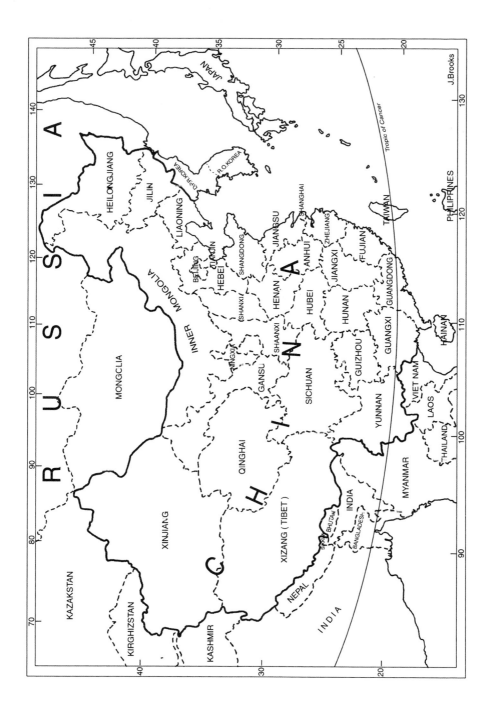

Printing: Mercedes-Druck, Berlin
Binding: Stein+Lehmann, Berlin